T0323527

Mobilisation of Forest Bioenergy in the Boreal and Temperate Biomes

Mobilisation of Forest Bioenergy in the Boreal and Temperate Biomes

Challenges, Opportunities and Case Studies

Evelyne Thiffault
Department of Wood and Forest Sciences and Research Centre on Renewable Materials, Laval University, Quebec, Canada

Göran Berndes
Physical Resource Theory, Chalmers University of Technology, Gothenburg, Sweden

Martin Junginger
Copernicus Institute, Utrecht University, Utrecht, The Netherlands

Jack N. Saddler
Forest Products Biotechnology/Bioenergy Group, University of British Columbia Vancouver, British Columbia, Canada

C.T. Smith
University of Toronto, Toronto, Canada

AMSTERDAM • BOSTON • HEIDELBERG • LONDON
NEW YORK • OXFORD • PARIS • SAN DIEGO
SAN FRANCISCO • SINGAPORE • SYDNEY • TOKYO
Academic Press is an imprint of Elsevier

Academic Press is an imprint of Elsevier
125 London Wall, London EC2Y 5AS, United Kingdom
525 B Street, Suite 1800, San Diego, CA 92101-4495, United States
50 Hampshire Street, 5th Floor, Cambridge, MA 02139, United States
The Boulevard, Langford Lane, Kidlington, Oxford OX5 1GB, UK

Notices
Knowledge and best practice in this field are constantly changing. As new research and experience
broaden our understanding, changes in research methods, professional practices, or medical
treatment may become necessary.

Practitioners and researchers must always rely on their own experience and knowledge in evaluating
and using any information, methods, compounds, or experiments described herein. In using such
information or methods they should be mindful of their own safety and the safety of others,
including parties for whom they have a professional responsibility.

To the fullest extent of the law, neither the Publisher nor the authors, contributors, or editors, assume
any liability for any injury and/or damage to persons or property as a matter of products liability,
negligence or otherwise, or from any use or operation of any methods, products, instructions, or
ideas contained in the material herein.

Library of Congress Cataloging-in-Publication Data
A catalog record for this book is available from the Library of Congress

British Library Cataloguing-in-Publication Data
A catalogue record for this book is available from the British Library

ISBN: 978-0-12-804514-5

For information on all Academic Press publications
visit our website at https://www.elsevier.com/

Working together
to grow libraries in
developing countries

www.elsevier.com • www.bookaid.org

Publisher: Joe Hayton
Acquisition Editor: Raquel Zanol
Editorial Project Manager: Mariana Kühl Leme
Editorial Project Manager Intern: Ana Claudia Garcia
Production Project Manager: Sruthi Satheesh
Designer: Greg Harris

Typeset by Thomson Digital

Contents

7. **Challenges and Opportunities for the Conversion Technologies Used to Make Forest Bioenergy**

William Cadham, J. Susan Van Dyk, J.S. Linoj Kumar, Jack N. Saddler

8. **Challenges and Opportunities for International Trade in Forest Biomass**

Patrick Lamers, Thuy Mai-Moulin, Martin Junginger

9. **Constraints and Success Factors for Woody Biomass Energy Systems in Two Countries with Minimal Bioenergy Sectors**
David C. Coote, Evelyne Thiffault, Mark Brown

10. **Challenges and Opportunities for the Mobilisation of Forest Bioenergy in the Boreal and Temperate Biomes**
Evelyne Thiffault, Göran Berndes, Patrick Lamers

List of Contributors

Numbers in parentheses indicate the pages on which the authors' contrbutions begin.

Antti Asikainen (10, 68), Natural Resources Institute Finland (LUKE), Joensuu, Finland

Göran Berndes (190), Physical Resource Theory, Chalmers University of Technology, Gothenburg, Sweden

Mark Brown (165), Australian Forest Operations Research Alliance, University of the Sunshine Coast, Sippy Downs, QLD, Australia

William Cadham (102), Forest Products Biotechnology/Bioenergy Group, University of British Columbia Vancouver, BC, Canada

David C. Coote (165), School of Ecosystem and Forest Sciences, University of Melbourne, Richmond, VIC; Australian Forest Operations Research Alliance, University of the Sunshine Coast, Sippy Downs, QLD, Australia

Guillaume Cyr (36), Canadian Forest Service, Natural Resources Canada, Quebec City, Canada

Ger Devlin (10), UCD Forestry, University College Dublin, Dublin, Ireland

Gustaf Egnell (50), Department of Forest Ecology and Management, Swedish University of Agricultural Sciences, Umeå, Sweden

Luc Guindon (36), Canadian Forest Service, Natural Resources Canada, Quebec City, Canada

Tanja Ikonen (68), Natural Resources Institute Finland (LUKE), Joensuu, Finland

Martin Junginger (127), Copernicus Institute, Utrecht University, Utrecht, The Netherlands

J.S. Linoj Kumar (102), Forest Products Biotechnology/Bioenergy Group, University of British Columbia Vancouver, BC, Canada

Patrick Lamers (50, 127, 190), Idaho National Laboratory, Idaho Falls, ID, United States of America

Thuy Mai-Moulin (127), Copernicus Institute, Utrecht University, Utrecht, The Netherlands

David Paré (36, 50), Canadian Forest Service, Natural Resources Canada, Quebec City, Canada

Johanna Routa (68), Natural Resources Institute Finland (LUKE), Joensuu, Finland

Jack N. Saddler (102), Forest Products Biotechnology/Bioenergy Group, University of British Columbia Vancouver, BC, Canada

Evelyne Thiffault (1, 10, 36, 50, 165, 190), Department of Wood and Forest Sciences and Research Centre on Renewable Materials, Laval University, Quebec City, Canada

J. Susan Van Dyk (102), Forest Products Biotechnology/Bioenergy Group, University of British Columbia Vancouver, BC, Canada

William A. White (84), Kingsmere Economics Consulting, Edmonton, AB, Canada

Preface

The International Energy Agency (IEA) Bioenergy aims to achieve a substantial bioenergy contribution to future global energy demands. Accelerating production and use of environmentally sound, socially accepted and cost-competitive bioenergy will help to provide increased security of supply, while reducing greenhouse gas emissions from energy use. Colleagues affiliated with several IEA Bioenergy Tasks have engaged to help achieve this objective, including Task 37 (Energy from Biogas), Task 38 (Climate Change Effects of Biomass and Bioenergy Systems), Task 39 (Commercialising Conventional and Advanced Liquid Biofuels from Biomass), Task 40 (Sustainable International Bioenergy Trade: Securing Supply and Demand), Task 42 (Biorefining—Sustainable Processing of Biomass into a Spectrum of Marketable Bio-based Products and Bioenergy) and Task 43 (Biomass Feedstocks for Energy Markets). This collaborative effort has led to the inter-Task project 'Mobilising Sustainable Bioenergy Supply Chains'. The purpose of this project is to identify sustainable biomass systems and promote their implementation. This was done by identifying a number of case studies that were evaluated in terms of potential bioenergy supply, barriers for development and potential policies to further mobilise the potential.

This book presents findings on forest biomass supply chains in the boreal and temperate biomes. It deals more specifically with countries from North America, the European Union and Oceania.

Work was coordinated by Evelyne Thiffault, assistant professor at the Department of Wood and Forest Sciences, Laval University (Canada) and a member of the Research Centre on Renewable Materials. The following scientific editors were also involved:

- Göran Berndes, Chalmers University of Technology, Sweden, IEA Bioenergy Task 43;
- Martin Junginger, Copernicus Institute, Utrecht University, the Netherlands, IEA Bioenergy Task 40;
- Jack N. Saddler, University of British Columbia, Canada, IEA Bioenergy Task 39;
- C.T. (Tat) Smith, University of Toronto, Canada, IEA Bioenergy Task 43.

Kirsten Hannam, from Agriculture and Agri-food Canada, acted as technical editor.

The following authors were involved:

- Antti Asikainen, Johanna Routa and Tanja Ikonen, LUKE, Finland;
- Mark Brown, Forest Industries Research Centre, University of Sunshine Coast, Australia;
- David Coote, Department of Forest and Ecosystem Science, The University of Melbourne, Australia;
- Ger Devlin, UCD Forestry, University College Dublin, Ireland;
- Gustaf Egnell, Department of Forest Ecology and Management, Swedish University of Agricultural Sciences, Umeå, Sweden;
- Thuy Mai-Moulin, Copernicus Institute, Utrecht University, the Netherlands;
- Patrick Lamers, Idaho National Laboratory, United States;
- David Paré, Natural Resources Canada—Canadian Forest Service, Canada;
- William Cadham, Susan Van Dyk and Linoj Kumar, Forest Products Biotechnology/Bioenergy Group, University of British Columbia, Canada;
- Bill White, Kingsmere Economics Consulting, Canada.

The following provided reviews of various chapters:

- Dan Neary, US Forest Service, US Department of Agriculture, United States;
- Annette Cowie, New South Wales Department of Primary Industries, Australia.

The following contributors generously provided data and insights about forest biomass deployment in their specific countries:

Jianbang Gan, Texas A&M University, United States;
Mohammad Ghaffariyan, University of the Sunshine Coast, Australia;
Bo Hektor, Swedish Bioenergy Association, Sweden;
Dirk Jaeger, University of Freiburg, Germany;
Søren Larsen, University of Copenhagen, Denmark;
Daniel Len, USDA Forest Service, United States;
Didier Marchal, Direction of Forest Resources, Wallonia Public Service, Belgium;
Rut Serra, Fédération québécoise des coopératives forestières, Canada;
Adam Sherman, Biomass Energy Resource Centre, United States;
Megan Smith, Ontario Ministry of Natural Resources, Canada;
Inge Stupak, University of Copenhagen, Denmark.

The team gratefully acknowledges IEA Bioenergy Executive Committee for providing funding to the project.

Executive Summary

INTRODUCTION

In 2012, the United Nations Secretary General set up an agenda on sustainable energy, for which one of the objectives was to double the share of sustainable renewable energy in the global energy mix by 2030, and triple the share of modern renewables to replace the use of traditional biomass. The International Renewable Energy Agency (IRENA) developed a Renewable Energy Roadmap (REMAP) to explore how this target could be put into practice. In the REMAP 2030, biomass could account for 60% of the total final renewable energy use in 2030, totalling 108 Exajoules (EJ), with applications in all sectors. Global biomass supply in 2030 is estimated to range from 97 to 147 EJ/year, with about 24–43 EJ coming from forestry. As another illustration of longer-term biomass demand for energy, the Intergovernmental Panel on Climate Change reviewed 164 long-term global energy scenarios and found bioenergy deployment levels in the year 2050 ranging from 80 to 150 EJ/year, for 440–600 ppm CO_2^{eq} concentration targets and from 118 to 190 EJ/year, for less than 440 ppm CO_2^{eq} concentration targets. To indicate magnitudes, 100 EJ roughly corresponds to 13.7×10^9 m^3 of wood (at 7.3 GJ/ m^3 of wood). In comparison, the total global annual quantity of wood removals only increased from 2.75×10^9 m^3 to around 3×10^9 m^3 from 1990 to 2011.

With their generally mature forestry sector, countries from the boreal and temperate biomes are expected to play an important role in the mobilisation of forest biomass for energy and contribute to reach the targets set by international agencies. In those countries, the models are mostly based on long-rotation forestry, which presents unique challenges and opportunities relative to short-rotation woody crops or forestry in tropical/sub-tropical areas.

AIM

The aim of the book is to identify opportunities and challenges for the mobilisation of sustainable forest bioenergy supply chains in the boreal and temperate biomes from North America, the European Union and Oceania. Secondary goals of the report include identifying the necessary elements of a successful and sustainable supply chain, as well as identifying occasions of knowledge transfer between countries and technologies.

ANALYSIS

A supply chain is a set of three or more entities (organisations or individuals) directly involved in the upstream and downstream flows of products, services, finances and information from a source to a customer. Unlike other manufacturing supply chains, forest product supply chains have a divergent co-production structure, in which trees are broken down into many products at all levels of the production processes. Wood has a highly heterogeneous nature, which makes planning and control a difficult task. Forest bioenergy supply, therefore, shares the inherent complexity of forest product supply chains in general.

Forest bioenergy supply chains evolve in specific geographical, socio-economical and policy environment and result in specific environmental and socio-economic footprints. The various processes, pathways and actors that comprise each forest bioenergy supply chain can present a range of opportunities and challenges, success stories, roadblocks and potential solutions in the development and deployment of sustainable business cases.

Comparison of Forest Biomass Supply Chains from the Boreal and Temperate Biomes

Several factors interact to determine the level of mobilisation of forest biomass supply chains, that is the ability of forest biomass to be harvested, collected, processed and delivered to end-users and markets in a manner that is competitive relative to other energy alternatives, notably to fossil fuels. Large differences exist between countries of the boreal and temperate biomes in terms of mobilisation, due to the action of individuals, businesses and organisations exploring opportunities that are shaped from real policy, market and operating conditions. Countries also vary in terms of the share of the forestry sector captured by forest energy. There is also variation in the integration (or lack thereof) of bioenergy within the basket of wood products in strategic, tactical and operational decision-making. Those differences represent a challenge when it comes to making useful recommendations that need to be meaningful and applicable throughout such an array of conditions. They also represent opportunities: these can be in the form of cross-regional and international synergies among stakeholders and markets with different but complementary characteristics; technological transfer; and possibilities of niche applications or efficiency gains arising from small improvements to various aspects of the supply chain, which can potentially result in large collective gains for global biomass mobilisation.

Quantifying Forest Biomass Mobilisation Potential in the Boreal and Temperate Biomes

Forest bioenergy mobilisation could be achieved via:
- Intensification of forest management activities, in which forestry would appropriate a larger share of forest ecosystem NPP; and
- Intensification of biomass recovery from silvicultural, harvesting and wood processing operations, in which bioenergy would appropriate a larger share of forest by-products/residues.

Mobilisation of forest biomass is relatively high in some countries (eg Belgium, Germany), while significant gains could be achieved in others (eg Canada, Russia). Forest biomass is likely to remain a commodity product that is strongly related to the management of the forest for other wood products. There is potential to increase biomass feedstock supply for bioenergy production in many regions, even without an enhancement of forest productivity by way of fertilisation, drainage or breeding improved tree species. For example, increasing the operational efficiency of logging residue recovery, that is the proportion of logging residues recovered from a given cutblock, represents one opportunity. Technological learning through improved worker training and changes in bioenergy policies and markets might increase this rate and, therefore, increase overall mobilisation of forest bioenergy.

On the other hand, intensification of forest management and biomass procurement can have adverse effects on biodiversity. For example, management of areas with high biodiversity value can be a concern. Strong environmental governance ensuring sustainable forest management practices and protection of forest ecosystem services is crucial.

In some regions, other constraints, such as access to the land, need also to be considered. In addition, certain regions are subjected to high rates of damage by natural disturbances, such as fire and insects. When it is not possible to reduce the damage from natural disturbances, the use of salvaged wood could be an important way to increase bioenergy mobilisation.

Environmental Sustainability Aspects of Forest Biomass Mobilisation

The environmental sustainability of forest biomass procurement needs to be well understood, as the capacity of ecosystems to provide biomass without negative impacts on ecological functioning limits the biomass potential. Emerging bioenergy markets typically first take advantage of secondary residue streams of various wood processing industries and tertiary end-of-life residues. The use of these secondary and tertiary wood resources is not likely to compromise the environmental sustainability of forests. When these resources in any region become scarce or fully utilised, primary residues (ie by-products of forest harvesting operations and silvicultural practices) such as branches, tops and non-merchantable trees become increasingly targeted as feedstock sources. Forest biomass procurement in the boreal and temperate biomes should, therefore, not be analysed as a stand-alone activity, but rather as an intensification of land use and of forest management, in which tree parts and trees are harvested in addition to conventional forest product fractions. Thus, principles of protection and sustainability should remain the same, whether forests are managed for conventional forest products only, or also for biomass for energy. Some modifications may be needed to find mitigation strategies for sensitive conditions where field evidence suggests that the incremental removal of biomass, or other forms of intensive management, may not be sustainable. Silvicultural practices such as fertilisation, competition control and soil preparation are options to manage the microenvironment and tree growing conditions, as well as preventing or mitigating negative impacts. Moreover, landscape management regulations should be put in place to ensure that sufficient biodiversity-important features, such as dead wood, aging stands, corridors etc. are preserved. Special attention should then be directed to trees and stands with high biodiversity values, or those important for maintaining ecosystem services. Applying the concept of adaptive forest management, ecological monitoring following harvesting operations, scientific field testing and modelling should be combined, in order to produce better knowledge that could help improve practices. The forestry sector needs to start adapting to a future situation where it is expected to provide conventional forest products, biomaterials and bioenergy. To achieve this, good governance mechanisms, such as landscape-level land-use planning and science-based improvements of practices, will become increasingly important to ensure sustainable forest product supply chains.

Challenges and Opportunities of Logistics and Economics of Forest Biomass Mobilisation

Mobilisation of forest biomass for energy production calls for a high level of integration with those forest industries that generate the raw materials (usually by-products of timber harvesting and wood processing). Integration of timber and energy supply chains helps overcome the challenge of seasonal fluctuations in energy demand by offering greater and more consistent use of machine capacity. Quality management of harvested biomass has become a very important issue in recent years. As supply volumes have increased, the economic losses associated with poor storage management have become obvious. Energy yields per unit of delivered biomass can be maximised through careful establishment and location of storage, prediction and measurement of changing moisture content and the ability to match supply with demand. Research shows that the use of biomass quality and location data to schedule wood chipping and transportation can markedly reduce the fleet size required to transport wood chips during periods of peak demand and the use of transport capacity through the year. Quality management becomes even more important when value-added products are made from biomass. The quality of liquid biofuels is highest when made from clean and dry feedstock. Producers of higher-value end products are willing to pay a premium price for good quality feedstock and their needs should be considered when handling forest biomass, in order to gain and maintain a competitive advantage. Biomass projects should target areas where market-driven competitiveness is best and where economic sustainability can be achieved with modest incentives.

Economic and Social Aspects of Forest Biomass Mobilisation

Sustainable resource use requires that environmental, social and economic sustainability be demonstrated; this holds true for bioenergy projects as well. The most common indicators of economic impact are employment and gross domestic product (GDP); these and other macroeconomic variables are usually influenced positively by the development of new bioenergy projects. The social impacts of bioenergy projects tend to be more varied and include land-use change, worker health and impacts on traditional communities, among others. Unlike the overwhelmingly positive impacts of bioenergy development on the economic side, social impacts are mixed. Barriers to increased bioenergy use include lack of accurate information, concerns over higher energy costs and anxiety about the potential for reduced government subsidies. Nevertheless, the use of bioenergy and other renewable energy sources can be encouraged by providing better information to consumers and producers, by improving analysis of potential social impacts before projects are put in place in order to prevent adverse effects that will hamper social acceptance in the future and by emphasising and encouraging the potential economic and environmental benefits of bioenergy.

Challenges and Opportunities for the Conversion Technologies Used to Make Forest Biomass-Based Bioenergy and Biofuels

Most of the forest biomass utilised around the world is employed to provide bioenergy, with the remainder being used to produce conventional forest products such as pulp and paper, sawnwood etc. Forest bioenergy is dominated by traditional, developing-world applications such as cooking and charcoal production, but developed-world applications, such as combined heat and power (CHP), pellet combustion etc., continue to expand. Modern bioenergy applications are expected to play an increasing role in the world's future renewable energy consumption. Forest biomass is already widely employed throughout the energy sector in heat/power generation, as part of stand-alone facilities, or integrated within industrial processes, residential heating (modern and traditional) and transportation. Although several conversion technologies, such as gasification, pyrolysis and cellulosic ethanol production, are being pursued, combustion will likely remain the most prevalent bioenergy process used to generate heat and power for the immediate future. Even after densification processes such as pelletisation or torrefaction, forest biomass has a low energy density relative to fossil fuels. Therefore, it is likely that most forest-derived biomass will continue to be used locally because transportation over long distances is economically challenging. However, in some cases, upgrading technologies such as pelletisation, pyrolysis, gasification and biochemical conversion can create sufficiently energy-dense, adaptable and carbon-friendly fuels to be attractive for import. The drivers of forest-based bioenergy consumption are not entirely based on direct cost comparisons with fossil fuels. Other priorities, such as climate change mitigation, energy security, GHG reduction, rural employment etc., also influence government support for bioenergy. The different approaches used to encourage bioenergy production and consumption in each country is a product of government policies, programmes and underlying political motivations. Modern biomass conversion technologies offer enormous opportunities in both developing and developed countries, through better use of forest- and mill-derived residues, improved efficiency of biomass conversions, reduced energy-related carbon emissions and enhanced energy security.

Challenges and Opportunities of International Trade of Forest Biomass

In an effort to reduce fossil fuel consumption, the use of woody biomass for heat and power generation is growing. Key destination markets will be countries within the European Union, particularly the United Kingdom, the Netherlands, Denmark and Belgium. While demand from Asia (particularly South Korea and Japan) will also increase, it will continue to play a secondary role. Across the European Union, adoption of sustainability criteria for forest biomass, based on the Renewable Energy Directive (RED) 2009/28/EC, or similar criteria, is likely. In the United Kingdom, the largest market for traded wood pellets, such criteria have already been proposed. As of 2015, only forest biomass that achieves at least a 60% reduction in GHG emissions relative to the EU fossil fuel electricity average can be used for bioenergy production and proof of sustainable forest management (SFM) is required. In the Netherlands, the energy industry and non-governmental organisations achieved principal agreement on sustainability criteria for solid biomass, but issues of compliance testing and monitoring must still be addressed. In Flemish Belgium, the sustainability requirements for bioliquids may soon be applied to woody biomass. In Denmark, a voluntary industry agreement is set to ensure that all bioenergy production by 2019 is conducted sustainably. At present, internationally traded wood pellets from temperate and boreal biomes originate primarily from the United States, Canada and Russia. Among these, Canada offers large stretches of SFM-certified forests, but Southeast United States has seen the strongest increase in wood pellet production and export in recent years. The expansion of pellet production in Southeast United States is mainly linked to available forest inventory (pulpwood in particular) and the competitive advantage gained by the relative proximity to demand markets in the European Union. In addition to the mobilisation barriers observed in supply chains at the local and national scale, limitations in the supply of forest biomass for international trade will be influenced by regulatory measures that could restrict the trade of specific feedstock fractions.

Constraints and Success Factors for Woody Biomass Energy Systems in Two Countries with Minimal Moody Biomass Energy Sectors

In many EU countries, ambitious targets have been set for reducing GHG emissions by increasing the contribution of woody biomass to renewable energy production. Although nations with active forest sectors could be subject to unique factors that discourage the increased use of wood for energy, lessons learned from nations with well-developed woody biomass energy sectors could be used to assist in the development of nascent woody biomass energy sectors elsewhere. The energy sectors of the EU nations exhibit striking differences with Australia and Canada. Of particular interest are critical differences in the supply of domestic primary energy, energy used per unit GDP, sovereign energy security, the mix of energy sources utilised, the economic viability of non-hydro renewables and the extent to which economies depend on fossil fuels. The policies affecting renewable energy sector development, including woody biomass availability and GHG reduction targets, also differ significantly. Adoption of new technologies with niche applications, leveraged with knowledge from the best practices and experience of countries with well-established bioenergy supply chains, represent an opportunity to set strong foundations for bioenergy production in countries with immature bioenergy sectors. It appears *prima facie* that the goal of reducing GHG emissions is only one of many factors that determine how much emphasis individual nations place on developing non-hydro renewable energy sectors (including woody biomass). The development of a woody biomass energy sector may be driven most strongly by regional issues, such as the need for less expensive waste disposal, cost-effective substitutions for fossil fuels, local economic development and energy resilience.

SUMMARY OF OPPORTUNITIES

Biomass production in forestry will have to increase drastically to support global bioenergy deployment targets. Forest biomass supply chains for bioenergy production can take on many forms, with variations in biomass source and management, end-products, conversion technology, logistics, environmental impacts and markets. Under the appropriate conditions, increasing bioenergy's share in the energy mix can contribute to important global targets, such as reducing GHG emissions, enhancing energy security and promoting sustainable development. However, bioenergy will bring value to society only if the benefits it provides exceed related externalities, as well as the opportunity costs of its development. Compared to other sources of renewable energy (and fossil fuels), bioenergy occupies a unique place due to its multisectoral nature and potential influence on environmental, social and economic conditions at various scales.

   As already experienced in several countries, readily accessible forest industry by-products such as bark, black liquor, sawdust and shavings are initially used as bioenergy feedstocks and there are few environmental concerns associated with this biomass use for energy. Logging residues, non-commercial roundwood and plantations represent complementary resources that can support ramping up to significantly larger scales if wood energy prices stimulate mobilisation. The impacts of utilising these resources are location specific and their availability across forest landscapes depends on both logistic factors and management/harvest guidelines safeguarding soils, water quality, biodiversity and other values. Other biomass resources, such as pulpwood quality logs, may also become used for energy, depending on the competitiveness of bioenergy compared to other feedstock uses.

Adjustments in forest management and harvesting regimes due to bioenergy demand will reflect a range of environmental, social and economic factors, including forest type, climate, legislation, other forest product markets, forest ownership and the character and product portfolio of the associated forest industry. For example, rising demand for forest bioenergy can cause competition for low-quality sawtimber and pulpwood, while, in other instances, adaptation of forest management planning to obtain an economically optimal output of forest products, placing equal weight on sawtimber, pulpwood and bioenergy feedstocks, can result in that both pulpwood and bioenergy feedstocks output increase, due to increased thinning frequency.

Thus, bioenergy demand can affect wood use in conventional forest products in antagonistic ways, for example when competition for the same feedstock drives up prices, impairing the competitiveness of conventional forest products. But it can also affect wood use in synergetic ways, where new opportunities and additional incomes from bioenergy use strengthen the forest industry. Besides the stated example where previously unused biomass (eg harvesting or processing residues) is used as bioenergy feedstock, synergies can also be realised through a change in forest management, when considering bioenergy market demand.

On the other hand, higher prices also provide incentives for expanding forests and improving forest productivity to support increased wood output. For example, the anticipation of emerging bioenergy markets can cause forest expansion (or reduce the rate of forest conversion to other land uses, such as urbanisation). As prices increase, forest management also intensifies and the amount of unmanaged forests being put under management rises. The character of the supply side response depends on price levels, but also on the pace of biomass demand growth. Forest planting and silvicultural measures to enhance forest growth represent slower supply side responses than the options to collect felling residues and utilise by-flows in the forest industry.

Global industrial roundwood prices have stayed close to or below 100 US\$/m^3 (in real US\$ of 1997) and energy wood prices around 50 US\$/m^3 (6.90 US\$/GJ) for most of the last 50 years. For their part, coal prices as of 2015 are around 3 US\$/GJ and they are expected to stay more or less constant, at least for the near future. It is difficult to imagine that where wood and coal compete for the same market, such as the energy sector (heat and power), the use of woody biomass becomes more attractive, unless the cost of using coal increases drastically. Thus, an increased bioenergy demand would require either subsidies for bioenergy, or policies that reflect the true costs of using fossil fuels. However, complementary and alternative strategies to such policies include identifying barriers, opportunities and solutions to mobilise feedstock and promote sustainable supply chains more efficiently.

Technological and Institutional Learning

The variability of forest biomass supply chains offers opportunities for further mobilisation by multiplying the occasions of learning. The following mechanisms of learning may all play a role in the various contexts of forest biomass mobilisation.

Learning-by-searching, that is improvements due to research, development and demonstration (RD&D), is the most dominant mechanism in the early stages and, to some extent, also during niche market deployment. Often also during the stages of pervasive diffusion and saturation, RD&D may contribute to technology enhancements.

Learning-by-doing comes from the repetition of production processes and leads to improvements, such as increased labour efficiency, work specialisation and production method improvements.

Learning-by-using comes from feedback from user experiences and can occur as soon as a technology is being used.

Learning-by-interacting is related to the increasing diffusion of the technology. During this process, the network interactions between actors such as research institutes, industry, end-users and policy makers generally improve and the above-mentioned mechanisms are reinforced.

Upsizing (or downsizing) a technology may lead to lower specific unit costs (eg the costs per unit of capacity).

Simple technology transfers from one supply chain to the other is not enough to create successful business cases. Technology and know-how need to be combined with existing expertise. An example of this would be economically successful business models that combine calculations of economically and technologically available resources made by experts, local knowledge of practitioners (eg effective work methods) and social innovations made by local entrepreneurs (eg forest bioenergy cooperatives).

Trade

Trade between countries within regions (eg European Union and North America) and between continents (eg Canada and the United Kingdom) offers opportunities and incentives for biomass mobilisation. So far, fuelwood, charcoal, wood chips and wood waste have almost exclusively been traded within regions; due to limited homogeneity and bulk density (eg fuelwood), high moisture content (eg wood chips), as well as a lack of handling equipment (eg in transloading stations). Forest biomass trade for energy between continents has so far only been economically viable for wood pellets, but other alternatives, such as torrefied wood and pyrolysis oil, may become increasingly traded if commercial production expands. If liquid biofuel production from wood becomes commercially available, it can be expected that these biofuels are traded between continents in the same way as ethanol and biodiesel are traded today. Trade can enable the establishment of logistic systems required for mobilisation of bioenergy supply chains in countries that currently do not have a large domestic market. For example, the current expansion of the US wood pellet production capacity, destined for export to the European Union, could provide a market and logistical 'stepping-stone' to the transition of the US feedstock supply system necessary to support a large national bio-economy.

Organisation Structures

Another opportunity to mobilisation that could rise from increased collaboration among stakeholders is the expansion of markets throughout organisation structures working with forest biomass, such as cooperatives, energy firms and trade centres. The open exchange of information, best practices and market instruments like long-term contracts could be used to improve cooperation between individual forest owners, forest owner organisations, entrepreneurs and forest industry to secure supply and demand. Well-functioning forest owner associations have proved their capability to increase wood supply from small-scale private properties. Rural development policies should continue to support capacity building of forest owner associations to encourage further mobilisation. Furthermore, support for organisation structures such as cooperatives (including items such as the development of professional corporations, associations and formal educational programmes) can also be a way to increase the professionalisation of the workforce in forest biomass supply chains. The establishment of a price index for bioenergy products, such as wood pellets, could also contribute to reducing short-term contracts, variable pricing and market instability.

Improvement of Supply Chain Data Reporting

There is a crucial need to analyse in greater detail the potential biomass supply, taking into account local conditions such as costs, ownerships patterns, quality requirements, infrastructure and environmental considerations. There is also the need for streamlining datasets on traded biomass volumes across international institutions. This would both improve reporting of trade streams and provide essential information for scientific research on mapping future trade streams, under different policy and potential trade regime scenarios.

Integration of Energy and Forest Systems

Management of Biomass Quality

One important step in forest biomass mobilisation is collaboration among stakeholders along the supply chain. This includes interactions to get a better understanding of needs in terms of feedstock and end-product characteristics. Technology developers and providers should ensure that the technologies developed are robust enough to handle the variability of forest biomass resources. On the other hand, biomass suppliers need to ensure strict feedstock quality management.

Integrated Planning of Bioenergy and Conventional Wood Products

Adequate characterisation and sorting of wood fibre as early as possible in the supply chain can provide strategic information facilitating economic and environmental management decisions on treatments for individual trees and forest stands, improve thinning and harvesting operations, as well as allocating timber resources efficiently for optimal utilisation. This should increase the profitability of the entire forest product value chain and the mobilisation of biomass as a result of two processes. First, proper identification, inventory and management of biomass for bioenergy (ie unutilised or unloved fibre by industries of conventional wood products) should increase the total volume of wood harvested per unit area of land and, thus, decrease overall harvesting costs. Second, the addition of bioenergy to the forest product value chain should improve the wood sorting capacity throughout the chain, thus ensuring that only feedstock of suitable quality is processed into each forest product. Optimal allocation of wood resources is vital, if wood and bioenergy buyers are to obtain the material most suited to their needs and suppliers are to obtain the best return for their investment in forest land.

An accompanying conclusion to this is that supply costs can be significantly reduced by integrating supplies of wood for both conventional forest products and bioenergy, rather than providing them via separate supply chains. Fibre terminals where wood can be sorted into multiple feedstock assortments and pre-processed (or blended) based on its characteristics can play a key role in the provision of such flexibility and links back to the above-mentioned organisation and logistic structures. For biomass producers, terminals could also ensure that forest machinery can be utilised effectively year-round. Since raw forest biomass cannot be transported over long distances, robust value-upgrading at terminals close to the feedstock sources, before long-distance transportation, could be considered.

The forest and the energy sectors use different types of measurements, which hinder communication and coordination between both sectors and market development. Forest industry and energy producers should work jointly on the interoperability of specifications and measures (volumetric and energetic) for wood fuel and wood fuel products, as well as common terminology and conversion factors related to wood for energy.

Conversion Efficiency and Cascading Use

Integration of forests and energy systems can help improve the efficiency in biomass resource use. A key is to strive for high efficiency in converting primary biomass into energy products and further into energy services. Higher value forest biomass resources can be used as feedstock several times through cascading systems where the biomass is processed into a material product, possibly multiple times (eg construction wood, followed by other material applications such as biochemicals and biomaterials) before it is finally used for energy.

Integrated Forest Land Planning for Energy, Conventional Wood Products and Ecosystem Services

There are many different silvicultural practices that could be modified/enhanced in such a way as to incorporate the future forest biomass market considerations at earlier planning stages. Along the same lines, forest management approaches aimed at producing forest bioenergy along with conventional forest products should focus on strengthening existing environmental synergies with other forest functions. These synergies could include forest fire protection, conservation of a balance in soil nutrients and support of biodiversity and water quality. This can be accomplished by increasing biomass removals in areas at high risk of forest fires in order to reduce fuel loading; by avoiding biomass procurement in sensitive forest areas characterised by poor soils; by leaving adequate amounts of dead wood and trees with cavities in forests in order to support biodiversity; by avoiding soil compaction by carrying out energy wood extraction at the same time as other forestry operation. A transition to an energy system devoid of fossil fuels and based on renewable sources (which may require short-term increases in GHG emissions) necessitates the adaptation of the forestry sector for a future situation where it is expected to provide ecosystem services, biomaterials and bioenergy.

Development of a Shared Vision

Recognition of Different Views and Understandings

While specific parts of the forest bioenergy supply chain may have the potential to expand rapidly, constraints related to social acceptability (such as evidence of activities becoming 'trusted' or 'taken for granted' by stakeholders in the general public and in markets) can slow or prevent their deployment. Such issues need to be recognised and must then be factored into longer-term plans for the development of the sector. The simple detection and formal recognition of different scientific understandings can contribute to produce new knowledge, shed new light on the environmental consequences of bioenergy policy and possibly bring stakeholders closer to a shared vision.

Development of Common Sustainability Criteria

Whether bioenergy development will be beneficial or detrimental for forests, as well as for people depending on forests for their livelihoods, will be determined by many factors, including legislation, regulations, standards and incentives for biomass production. Development of sustainability criteria for bioenergy is part of the shared vision. Stakeholders along forest biomass supply chains recognise clearly that there is a need to substantiate the sustainable production of biomass. On the other hand, consensus on what should be considered sustainable production and how this should be implemented, certified and monitored, is still elusive. A dialogue aiming to establish internationally accepted sustainability requirements for bioenergy commodities will create new opportunities for sustainable bioenergy production and trade.

Development of Common Technical Standards

Also part of the need for a shared vision is the drive towards technical standardisation of bioenergy products. The development of standards (such as the mandates given by the European Commission to the European Committee for Standardisation—CEN) can help remove trade barriers, increase market transparency and contribute to public acceptance. The fact that the major producing regions have already started to compare (and possibly align) their technical standards is a sign that international cooperation may lead to new opportunities for international bioenergy trade.

CONCLUSIONS

Forest biomass supply chains cover a wide range of biomass sources, logistic systems, conversion technologies, end-products and stakeholders. Several supply chains have proven to be economically viable and are considered sustainable; others not. Bioenergy-related policies should be designed in a way that they enhance technological and economic efficiency and environmental sustainability. Objectives and policies for bioenergy and for renewable energy in general are often formulated and agreed upon at higher decision making levels. However, the design of objectives and policies can be better informed by knowledge and experience at the lower levels of decision making, where the implementation takes place, in order to integrate effectively renewable energy into the existing energy systems. Local strategies can assist in translating national plans to local level action, while allowing for local level prioritisation and ownership. Local planning can facilitate the identification of the most favourable sites and technologies, improve the understanding of the local environment and its actors and facilitate the integration of policies throughout various sectors to cater for regional ecological complexity.

Finally, national governments around the world (besides those in the boreal and temperate biomes) have different reasons (policy objectives), as well as different available resources (including climate and available land) for increasing the production and use of forest biomass. Thus, an essential first step in designing appropriate bioenergy policies is to distinguish between the needs and resources of individual countries. Industrialised countries and major exporters of bioenergy should encourage the development of bioenergy where it can be demonstrated that doing so will reduce GHG emissions over the whole life-cycle. Developing countries and those with economies in transition should primarily develop bioenergy to benefit local livelihoods through providing heat and electricity, as well as affordable, safe and more efficient fuels and so support wider sustainable development goals without jeopardising food security. Ultimately, a multifaceted policy effort to support a broad array of technological, structural and behavioural changes are needed that enable energy and industrial transitions and development of supply chains that deliver feedstock to a range of conversion and utilisation routes.

Chapter 1

Introduction

Evelyne Thiffault
Department of Wood and Forest Sciences and Research Centre on Renewable Materials,
Laval University, Quebec City, Canada

Highlights

- Several international organisations have set up renewable energy targets for the coming decades; these targets include energy from forest biomass.
- Forest biomass could be used to meet a significant share of the growing global demand for biomass-based energy. Forests in the temperate and boreal biomes typically support well-developed forest sectors, are managed over long rotations and will have an important role to play in the mobilisation of forest bioenergy supply chains in the near future.
- As is the case for conventional forest products, forest bioenergy feedstocks flow along the supply chain, from forest contractors to production facilities, then through channels of distributors and wholesalers, to reach final end-users.
- Each step along the supply chain has an influence on the opportunities and barriers to the development and deployment of sustainable forest bioenergy business cases.

INTRODUCTION

In order to be considered renewable, energy must be derived from natural sources and be replenished more quickly than it is consumed. Biomass, including solid wood, plant and animal products, as well as gases and liquids derived from waste, can be considered a renewable source of energy. In its Global Action Agenda, the United Nations' Secretary General proposed that the contribution of renewables to global energy production be doubled in 2030, relative to 2010 figures, and that traditional uses of biomass, for example wood fuel and manure for cooking and heating in developing countries, be replaced with modern, efficient renewable energy technologies (Fig. 1.1) (SEALL, 2012). The International Renewable Energy Agency (IRENA) has developed a Renewable Energy Roadmap (REmap 2030), which outlines the plan to 2030 for working towards these goals (IRENA, 2014a). According to REmap 2030, modern bioenergy technologies could account for 60% of total final renewable energy use, totalling 108 exajoules (EJ) across all sectors (Fig. 1.2). By 2030, an estimated

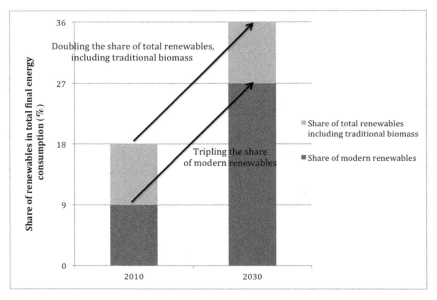

FIGURE 1.1   Target for renewable energy in REmap 2030 (IRENA, 2014a).

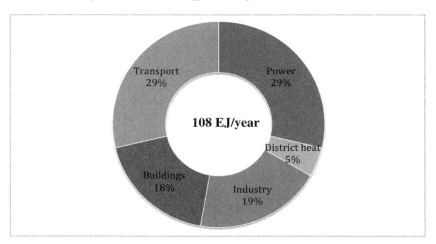

FIGURE 1.2   Biomass demand, by sector, in Remap 2030 (IRENA, 2014b).

97–147 EJ could be obtained from biomass each year; of this, 24–43 EJ could come from forests (IRENA, 2014b).

Similarly, the International Energy Agency (IEA) developed a plan (Blue Map 2050) for achieving the CO_2 emission reductions that are required by 2050 to limit the increase in global temperatures to less than 2°C (IEA, 2010). As part of this plan, the IEA estimates that biomass should provide close to 146 EJ by 2050 (Fig. 1.3). Moreover, a recent review of greenhouse gas (GHG) stabilisation scenarios, conducted by the Intergovernmental Panel on Climate Change

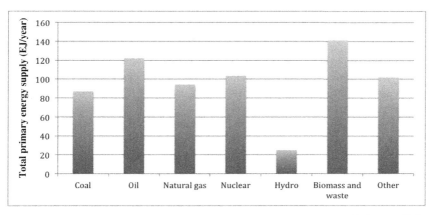

FIGURE 1.3   Total primary energy supply, by source, in the IEA BLUE Map 2050 (IEA, 2010).

(IPCC, 2011), indicated that bioenergy could contribute between 120–155 EJ and 265–300 EJ by 2050; by that time, the maximum technical potential of forest biomass (meaning the maximum biomass production potential, including technical barriers) could reach 110 EJ, although global estimates vary widely. Thus, forest bioenergy feedstocks (Box 1.1) could play a significant role in

---

BOX 1.1 Forest Biomass and Forest Bioenergy Feedstocks

In its literal definition, forest biomass includes all organic material found in a forest, both above and below ground, and both living and dead. This includes trees, crops, grasses, surface litter, roots, etc. When collected and used to produce energy, this material is referred to as forest bioenergy feedstock.

The definition of forest bioenergy feedstock used in this and subsequent chapters includes:

1. roundwood, including:
   a. roundwood currently harvested from the forest and used to produce conventional wood products, such as sawnwood, pulp and paper and fibreboard; and
   b. surplus forest growth, that is roundwood that could potentially be harvested over and above current harvesting rates while still remaining within the sustainable harvest rate of the forest;
2. primary forest residues, which are by-products of harvesting and silvicultural operations, such as tree tops, branches, prunings, small trees from early thinnings, etc. and
3. secondary forest residues, which are by-products of the industrial processing of wood, such as sawdust, wood shavings and wood chips.

Tertiary residues, such as demolition wood and non-recycled paper, are another potential source of bioenergy feedstocks, but they are not discussed in detail in this book.

modern bioenergy production and, indeed, in total global energy production in the future.

With their relatively mature forest sectors, countries from the boreal and temperate biomes (Fig. 1.4) will probably play an important role in the mobilisation of forest biomass for the modern production of energy (Fig. 1.5). These forests are typically managed over long rotations, that is the time between stand establishment and final cut is at least 20 years (Egnell and Björheden, 2013). Long-rotation forestry presents unique challenges and opportunities relative to short-rotation woody crops, or shorter-rotation forestry in tropical/sub-tropical areas.

The maturity of forest bioenergy supply chains varies widely among countries of the boreal and temperate biomes. In Finland and Sweden, forest bioenergy has been an integral part of the forest and energy portfolios for several decades, and numerous support policies have been implemented to ensure the stability of forest biomass supply chains. In Canada, by contrast, forest bioenergy represents only a small share of total energy production, and many communities, investors and policymakers remain sceptical about the viability of forest biomass-based business models. Therefore, tailored solutions are required to enhance forest biomass mobilisation, and establish or maintain stable forest biomass supply chains in different countries.

GENERAL AIM OF THE BOOK

The aim of this book is to identify the opportunities and challenges presented by the mobilisation of sustainable forest bioenergy supply chains in countries across the boreal and temperate biomes, in order to assist efforts to meet global targets for bioenergy.

At this point, it would be useful to define a supply chain. Mentzer et al. (2001) define it as a set of three or more entities (organisations or individuals) directly involved in the upstream and downstream flows of products, services, finances and information from a source to a customer (Fig. 1.6).

However, unlike other manufacturing supply chains, which have a convergent product structure, a typical forest product supply chain has a divergent co-production structure: trees are broken down into many products, such as sawnwood, chips for pulp, etc. at all levels of the production processes (Fig. 1.7). Moreover, wood and wood fibre have a highly heterogeneous nature, which makes planning and control a difficult task with regard to production output control (D'Amours et al., 2008; Frayret et al., 2007). Forest bioenergy supply therefore shares the inherent complexity of forest product supply chains in general (Shabani et al., 2013).

A forest bioenergy supply chain consists of discrete processes and a natural hierarchy of decision-making, including: strategic (long-term), tactical (medium-term) and operational (short-term) decisions. It also has a structure, which defines the connections between the facilities that work together to supply

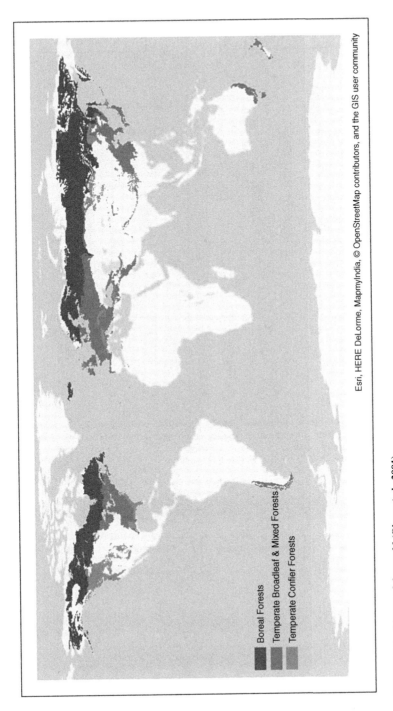

Esri, HERE DeLorme, MapmyIndia, © OpenStreetMap contributors, and the GIS user community

Boreal Forests
Temperate Broadleaf & Mixed Forests
Temperate Conifer Forests

FIGURE 1.4   Biomes of the world (Olson et al., 2001).

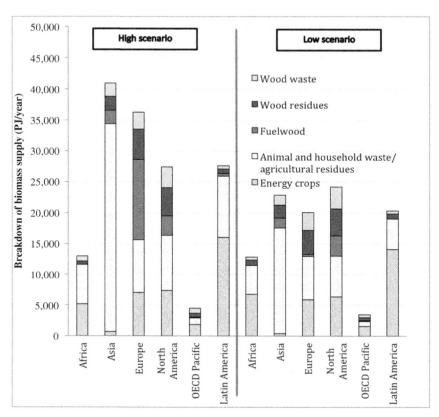

FIGURE 1.5 **Breakdown of biomass supply, by source and region, in the REmap 2030 (IRENA, 2014b).** Note: Biomass from forestry is included in the fuelwood, wood residue and wood waste categories.

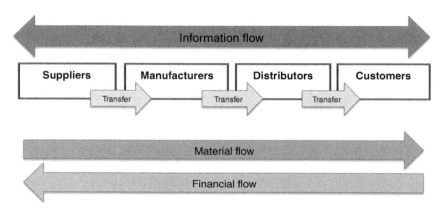

FIGURE 1.6 **Generic model of a supply chain (APICS, 2011).**

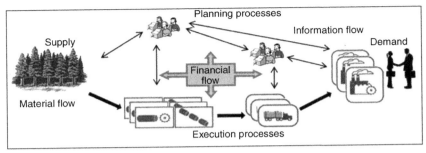

FIGURE 1.7   Forest product supply chain. *(Source: FORAC research consortium, Laval University.)*

a product or service, along which material, information and finances flow (Sharma et al., 2013). The supply chain also evolves in a given geographical, socio-economic and policy environment and results in a specific environmental and socio-economic footprint (Fig. 1.8).

The various processes, pathways and actors that comprise each forest bio-energy supply chain can present various opportunities and challenges, success stories, roadblocks and potential solutions in the development and deployment of sustainable forest bioenergy business cases. This book will discuss the mobilisation of sustainable forest bioenergy supply chains, using two approaches: first, by analysing the components of supply chains; and second, by comparing

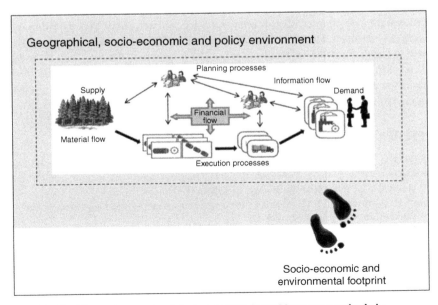

FIGURE 1.8   The environment and footprint of the forest bioenergy supply chain.

countries of the boreal and temperate biomes that have contrasting histories, policies and levels of forest bioenergy deployment. After describing typical forest bioenergy supply chains in various countries within the boreal and temperate biomes, as well as their specific environments (see chapter: Comparison of Forest Biomass Supply Chains From the Boreal and Temperate Biomes), subsequent chapters will discuss:

- the availability of forest bioenergy feedstocks (see chapter: Quantifying Forest Biomass Mobilisation Potential in the Boreal and Temperate Biomes);
- the environmental sustainability of forest biomass mobilisation (see chapter: Environmental Sustainability Aspects of Forest Biomass Mobilisation);
- the logistics and economic sustainability of forest biomass mobilisation (see chapter: Challenges and Opportunities of Logistics and Economics of Forest Biomass);
- the social and economic implications of forest biomass mobilisation (see chapter: Economic and Social Barriers Affecting Forest Bioenergy Mobilisation: A Review of the Literature); and
- the conversion processes and technologies involved in forest bioenergy production (see chapter: Challenges and Opportunities for the Conversion Technologies Used to Make Forest Bioenergy).

While forest biomass supply chains can be described in their local or regional context, they can also be analysed at a macro-scale, for example the international trade of forest bioenergy feedstocks (see chapter: Challenges and Opportunities for International Trade in Forest Biomass). Insights can also be gained by comparing countries at different stages of domestic forest bioenergy deployment (see chapter: Constraints and Success Factors for Woody Biomass Energy Systems in Two Countries With Minimal Bioenergy Sectors). Finally, lessons learnt from the information provided in previous chapters can be used to develop recommendations for addressing future challenges in forest biomass (see chapter: Challenges and Opportunities for the Mobilisation of Forest Bioenergy in the Boreal and Temperate Biomes).

REFERENCES

APICS, 2011. Certified Supply Chain Professional Learning System. APICS, Chicago, IL, Version 2.2.

D'Amours, S., Rönnqvist, M., Weintraub, A., 2008. Using operational research for supply chain planning in the forest products industry. INFOR 46 (4), 265–281.

Egnell, G., Björheden, R., 2013. Options for increasing biomass output from long-rotation forestry. WIREs Energy Environ. 2 (4), 465–472.

Frayret, J.-M., D'Amours, S., Rousseau, A., Harvey, S., Gaudreault, J., 2007. Agent-based supply-chain planning in the forest products industry. Int. J. Flexible Manufact. Syst. 19 (4), 358–391.

IEA, 2010. Energy Technology Perspectives 2010 Scenarios and Strategies to 2050. Organisation for Economic Co-operation and Development, Paris, Available from: https://www.iea.org/publications/freepublications/publication/etp2010.pdf.

IPCC, 2011. Special Report on Renewable Energy Sources and Climate Change Mitigation. Cambridge University Press, United Kingdom and New York, NY.

IRENA, 2014a. REmap 2030 A Renewable Energy Roadmap. IRENA, Abu Dhabi, Report available from: http://www.irena.org/remap.

IRENA, 2014b. Global Bioenergy Supply and Demand Projections: A Working Paper for Remap 2030. IRENA, Abu Dhabi, Available from: https://www.irena.org/remap/IRENA_REmap_2030_Biomass_paper_2014.pdf.

Mentzer, J.T., DeWitt, W., Keebler, J.S., Min, S., Nix, N.W., Smith, C.D., Zacharia, Z.G., 2001. Defining supply chain management. J. Bus. Logist. 22 (2), 1–25.

Olson, D.M., Dinerstein, E., Wikramanayake, E.D., Burgess, N.D., Powell, G.V.N., Underwood, E.C., D'Amico, J.A., Itoua, I., Strand, H.E., Morrison, J.C., Loucks, C.J., Allnutt, T.F., Ricketts, T.H., Kura, Y., Lamoreux, J.F., Wettengel, W.W., Hedao, P., Kassem, K.R., 2001. Terrestrial ecoregions of the world: a new map of life on Earth. Bioscience 51 (11), 933–938.

SEALL, 2012. A Global Action Agenda: Pathways for Concerted Action Towards Sustainable Energy for All. SEALL, New York, Available from: http://www.un.org/wcm/webdav/site/sustainableenergyforall/shared/Documents/SEFA-ActionAgenda-Final.pdf.

Shabani, N., Akhtari, S., Sowlati, T., 2013. Value chain optimization of forest biomass for bioenergy production: a review. Renew. Sustain. Energy Rev. 23, 299–311.

Sharma, B., Ingalls, R., Jones, C., Khanchi, A., 2013. Biomass supply chain design and analysis: basis, overview, modeling, challenges, and future. Renew. Sustain. Energy Rev. 24, 608–627.

Chapter 2

Comparison of Forest Biomass Supply Chains From the Boreal and Temperate Biomes

Evelyne Thiffault*, Antti Asikainen**, Ger Devlin[†]

*Department of Wood and Forest Sciences and Research Centre on Renewable Materials, Laval University, Quebec City, Canada; **Natural Resources Institute Finland (LUKE), Joensuu, Finland; [†]UCD Forestry, University College Dublin, Dublin, Ireland

Highlights

- There is wide variation among countries of the boreal and temperate biomes in terms of the organisation and maturity of forest bioenergy markets and the share of the forest sector occupied by bioenergy.
- The ability to make meaningful, applicable recommendations for mobilising forest biomass is challenged by differing realities among countries.
- Nevertheless, these differences can also represent opportunities, for example via technology transfer, which could yield large collective gains in global biomass mobilisation.

With contributions from:

Jianbang Gan, Texas A&M University, College Station, TX, United States

Mohammad Ghaffariyan, University of the Sunshine Coast, Sippy Downs, QLD, Australia

Bo Hektor, Swedish Bioenergy Association, Stockholm, Sweden

Dirk Jaeger, University of Freiburg, Freiburg, Germany

Daniel Len, USDA Forest Service, Atlanta, GA, United States

Didier Marchal, Wildlife and Forestry Department, Namur, Belgium

Rut Serra, Fédération québécoise des coopératives forestières, Quebec, Canada

Adam Sherman, Biomass Energy Resource Centre, Burlington, VT, United States

Megan Smith, Ontario Ministry of Natural Resources, Sault Ste. Marie, ON, Canada

Rien Visser, University of Canterbury, Christchurch, New Zealand

Mobilisation of Forest Bioenergy in the Boreal and Temperate Biomes

INTRODUCTION

The level of mobilisation of forest biomass supply chains varies widely among countries of the boreal and temperate biomes and depends on the ability of forest biomass to be harvested, collected, processed and delivered to end-users and markets, in a manner that is competitive with other energy sources. Forest biomass supply chains can vary in response to both supply- and demand-related factors. Jessup and Walkiewicz (2013) offer the following framework for identifying the factors driving the development of forest biomass supply chains:

Supply

- Forest and environmental conditions:
  - species mix;
  - climatic conditions (rainfall, soil conditions, growing season); and
  - terrain (steep, flat, soil types);
- Cultural influences;
- forest ownership/stewardship patterns;
- forest management goals and objectives (wood supply, sustainability, biodiversity, ecosystem management, conservation);
- economic health;
- financial health of forest sector (logging firms, transport, processing) and the ability to invest in new technology through equipment purchases; and
- Government policies and support (subsidies/production incentives).

Demand

- End-use markets:
  - direct heating (industrial/residential);
  - combined heat and power (CHP);
  - co-firing electrical plants; and
  - biofuels and biochemical;
- Government policies and intervention:
  - environmental mandates; and
  - feed-in tariffs;
- price and availability of competing energy (eg fossil fuels);
- other renewable energy sources;
- conventional roundwood and wood fibre markets for non-energy purposes; and
- technology in various transformation processes.

Within each country, the mobilisation of forest biomass is affected by individuals, businesses and organisations who are interested in exploring economic opportunities and who must operate within the constraints of policy, market and technological realities (Jessup and Walkiewicz, 2013). This chapter will provide an overview of the current context and practices of forest biomass supply chains in several countries in the boreal and temperate biomes, in order to set the stage for deeper analyses of opportunities and constraints to mobilisation in subsequent chapters. For this exercise, 12 countries with contrasting forest biomass supply chains were selected: Australia, Belgium, Canada, Croatia, Denmark, Finland, Germany, Ireland, New Zealand, Norway, Sweden and the United States. This chapter describes typical forest biomass supply chains in each country and discusses the geographic, socio-economic and forest management contexts that have shaped their development.

COUNTRY PROFILES

Geographical and Socio-Economic Characteristics

Among the countries discussed, Canada is the largest, followed by the United States and Australia; Belgium and Denmark are the smallest (Table 2.1). Within each country, the area of forested land varies from 548,000 ha (Denmark) to 310 million ha (Canada), and the share of forested land relative to the total landbase ranges from 11% (Ireland) to 73% (Finland) (Table 2.1). By contrast, the share of agricultural land is highest in Ireland and lowest in Canada, Sweden and Finland.

TABLE 2.1 Land Area and Population in 2012 (FAO, 2015)

Countries	Total land area (1000 ha)	Forest area (1000 ha)	Agricultural area (1000 ha)	Population (1000 inhabitants)	Population density (inhabitants/ km^2 of land area)
Australia	768,230	147,452	405,474	22,684	3
Belgium	3,028	680	1,333	11,060	365
Canada	909,351	310,134	65,346	34,880	4
Croatia	5,659	1,923	1,326	4,307	76
Denmark	4,243	548	2,624	5,598	132
Finland	30,389	22,157	2,285	5,414	18
Germany	34,854	11,076	16,664	81,932	235
Ireland	6,889	757	4,533	4,587	67
New Zealand	26,331	11,280	8,252	4,565	17
Norway	36,527	10,218	992	4,994	14
Sweden	40,734	28,203	3,049	9,519	23
United States	914,742	304,788	408,707	313,914	34

TABLE 2.2 Average GDP Per Capita (2000–2012) (FAO, 2015)

Countries	GDP per capita (US$ per capita)
Australia	36,351
Belgium	37,502
Canada	35,324
Croatia	10,840
Denmark	49,191
Finland	32,741
Germany	33,273
Ireland	38,482
New Zealand	26,568
Norway	70,786
Sweden	35,133
United States	44,379

The United States has the largest population (313 million people in 2012), while Croatia, New Zealand and Ireland have the smallest. However, population density is lowest in Canada and Australia (3–4 inhabitants/km^2) and highest in Germany (368 inhabitants/km^2) (Table 2.1). As an indicator of relative wealth, Norway and Denmark had the highest average gross domestic product (GDP) per capita between 2000 and 2012, followed by the United States, while Croatia had the lowest (Table 2.2).

Energy Profile

The United States has the highest total primary energy supply [89,281 petajoules (PJ) in 2012] of all the countries discussed; in fact, the United States has a greater energy supply than that of the other countries combined (Table 2.3). On a per capita basis, Canada and the United States have the highest total primary energy supply, while Croatia, Ireland and Denmark have the lowest (Table 2.3). In terms of the share of total primary energy that is supplied by renewables, Norway, New Zealand and Sweden ranked the highest in 2012, with 47, 37 and 36%, respectively; Australia, Belgium, Ireland and the United States ranked the lowest (Fig. 2.1). Nevertheless, most of the countries discussed in this analysis saw their share of renewable energy increase between 2002 and 2012. It is also noteworthy that several of these countries are important fossil fuel producers: Canada and Norway are net oil exporters and Australia, Canada, Germany and the United States extract large volumes of coal (see the discussion on the fossil fuel industry in chapter: Constraints and Success Factors for Woody Biomass Energy Systems in Two Countries With Minimal Bioenergy Sectors).

TABLE 2.3 Total Primary Energy Supply in 2012 (IEA, 2015)

Countries	PJ	PJ/1000 inhabitants
Australia	5,597	0.25
Belgium	2,343	0.21
Canada	10,578	0.30
Croatia	332	0.08
Denmark	726	0.13
Finland	1,402	0.26
Germany	12,869	0.16
Ireland	559	0.12
New Zealand	794	0.17
Norway	1,222	0.24
Sweden	2,046	0.21
United States	89,281	0.28

Note: Total primary energy supply is made up of production + imports – exports – international marine bunkers – international aviation bunkers ± stock changes.

Of the countries considered in this analysis, the United States obtains the highest total value of energy from liquid and solid biofuels and waste, as well as from primary solid biofuels alone, followed by Germany (Table 2.4). However, solid biofuels contribute the greatest share of total primary energy supply in Finland, Sweden and Denmark (24, 20 and 14%, respectively, in 2012). In the remaining countries, solid biofuels contribute approximately 5% of the total primary energy.

Forest Sector

Of the countries included in this analysis (except Australia, for which data was unavailable), Canada and the United States have the highest total value of forest growing stock. Sweden ranks the highest among European countries (Table 2.5). Relative to the size of the forest landbase, however, New Zealand and Belgium have the greatest density of growing stock per hectare of forest, while Canada, with its large areas of natural boreal forest, has the lowest (Table 2.5). Canada's forest landbase is dominated by primary forests [according to the United Nations' Food and Agriculture Organization (FAO) definition (FAO, 2010)], while more than 75% of the forests in Ireland and Denmark have been planted (Fig. 2.2). Most of the forests in Europe and the United States are privately owned, but the majority of forests in Australia, Canada, Croatia, Ireland and New Zealand are under public ownership (Fig. 2.3).

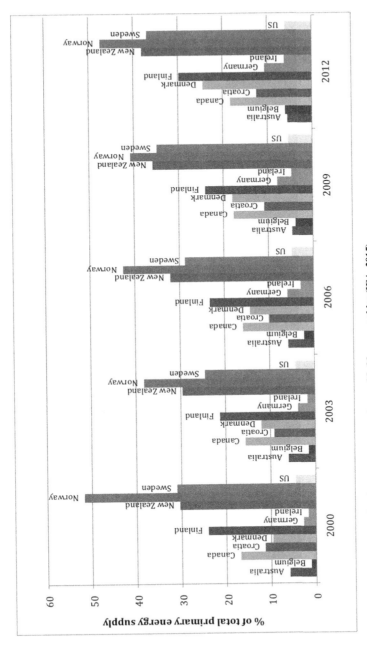

FIGURE 2.1 Share of total primary energy supply provided by renewables (IEA, 2015).

TABLE 2.4 Primary Energy Supply From Biofuels and Waste in 2012 (IEA, 2015)

Countries	Biofuels and waste (PJ)	Primary solid biofuels only (PJ)
Australia	212	177
Belgium	223	83
Canada	519	425
Croatia	23	20
Denmark	155	104
Finland	361	332
Germany	1171	494
Ireland	18	9
New Zealand	50	47
Norway	72	49
Sweden	488	400
United States	3710	1866

Note: Biofuels and waste include solid biofuels, biogases, liquid biofuels and waste. Primary solid biofuels are defined as any plant matter used directly as fuel, or converted into other forms, before combustion. This covers a multitude of woody materials generated by industrial processing, or provided directly by forestry and agriculture (fuelwood, wood chips, bark, sawdust, shavings, chips, black liquor, animal materials/wastes and other solid biofuels). Note that, for biofuels, only the amounts of biomass specifically used for energy purposes are included in the energy statistics. Therefore, the non-energy use of biomass is not taken into consideration and is null, by definition.

TABLE 2.5 Forest Growing Stock in 2010 (FAO, 2010)

Countries	Total growing stock (Mm^3 over bark)	Growing stock per forest land area (m^3/ha of forest land)
Australia	n/a	n/a
Belgium	336	494
Canada	32,983	106
Croatia	831	431
Denmark	219	400
Finland	4,360	197
Germany	3,492	315
Ireland	147	194
New Zealand	4,362	529
Norway	2,024	198
Sweden	6,739	239
United States	90,180	296

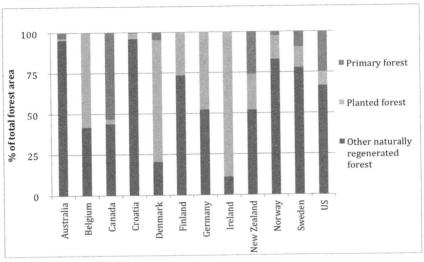

FIGURE 2.2   Primary, planted and other naturally regenerated forests in 2010 (FAO, 2010).

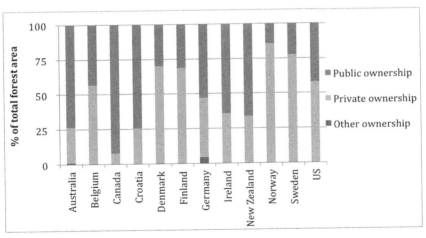

FIGURE 2.3   Structure of forest ownership in 2010 (FAO, 2010).

Of the countries included in this analysis, the United States is the biggest producer of roundwood (including both industrial roundwood and wood used as fuel), with an average annual production of over 360 millions m^3/year between 2000 and 2012, followed by Canada (Table 2.6). Of the European countries, Sweden, Germany and Finland are the biggest producers of roundwood. Relative to their respective forest areas, Belgium, Denmark and Germany harvest the most roundwood volume per hectare of forest each year, while Australia and

TABLE 2.6 Average Roundwood Production (2000–2012) (FAO, 2015)

Countries	Roundwood (m^3/year)	Roundwood (m^3/year per ha of forest land)	Industrial roundwood (m^3/year)	Fuelwood (m^3/year)
Australia	31,187,615	0.21	25,752,923	5,434,692
Belgium	4,773,725	7.02	4,098,030	675,695
Canada	170,696,704	0.55	167,959,487	2,737,217
Croatia	4,254,308	2.21	3,280,846	973,462
Denmark	2,333,751	4.26	1,366,831	966,920
Finland	51,886,499	2.34	46,750,464	5,136,035
Germany	54,120,102	4.89	47,057,871	7,062,232
Ireland	2,580,234	3.41	2,499,062	81,173
New Zealand	21,567,462	1.91	21,567,462	n/a
Norway	9,486,394	0.93	7,805,344	1,681,050
Sweden	70,615,308	2.50	64,715,308	5,900,000
United States	422,746,186	1.39	379,522,692	43,223,494

Note: Roundwood includes all roundwood felled or otherwise harvested and removed. It comprises all wood obtained from removals, that is the quantities removed from forests and from trees outside the forest, including wood recovered from natural, felling and logging losses during the period, calendar year or forest year. It includes all wood removed with or without bark, including wood removed in its round form, or split, roughly squared, or in other form (eg branches, roots, stumps and burls—where these are harvested—and wood that is roughly shaped or pointed. It is an aggregate comprising fuelwood, including wood for charcoal and industrial roundwood. It is reported in cubic metres solid volume underbark (ie excluding bark). Industrial roundwood is an aggregate comprising sawlogs and veneer logs, pulpwood (round and split) and other industrial roundwood. Fuelwood includes roundwood that will be used as fuel for purposes such as cooking, heating or power production. It includes wood harvested from main stems, branches and other parts of trees (where these are harvested for fuel) and wood that will be used for the production of charcoal, wood pellets, etc. It also includes wood chips that are made directly (ie in the forest) from roundwood.

Canada harvest the least. Denmark and Croatia use the greatest proportion of harvested roundwood for fuel (Table 2.6).

Forest products account for the greatest proportion of total national exports in Finland, followed by Sweden, New Zealand, Croatia and Canada (Table 2.7). In absolute terms, however, Canada exports the greatest value (in US$) of forest products to the rest of the world, followed by the United States, Germany and Sweden. At the same time, the United States imports the greatest value of forest products from the rest of the world, followed by Germany. Canada, Sweden and Finland have large positive trade balances in terms of forest product import/export value (ie they export more forest products than they import). Croatia, Germany New Zealand and Norway also have positive trade balances, while the remaining countries have negative trade balances (Table 2.7).

TABLE 2.7 Forest Product Exports and Imports (FAO, 2014, 2015)

Country	Exports of forest products (% of national exports) 2011	Exports (1000 US$) 2000–2012 average	Imports (1000 US$) 2000–2012 average
Australia	0.9	1,491,411	1,922,238
Belgium	2.2	4,997,602	5,431,235
Canada	6.2	24,729,275	4,679,159
Croatia	7.3	417,570	386,891
Denmark	3.1	502,020	2,138,795
Finland	19.8	12,782,862	1,632,246
Germany	2.9	17,236,144	15,798,735
Ireland	0.6	458,267	859,038
New Zealand	9.5	2,102,647	440,567
Norway	1.2	1,741,522	1,262,438
Sweden	11.2	13,560,812	2,347,161
United States	2.5	19,102,710	24,843,154

Production of primary goods provides the greatest number of forest sector jobs in the United States, Canada and Germany (Table 2.8). Jessup and Walkiewicz (2013) suggested that the amount of roundwood removals per forest sector employee can be used as an indicator of the degree of mechanisation in the forest industry. According to this index, mechanisation is highest in New Zealand and lowest in Denmark and Croatia (Table 2.8).

In summary, the countries discussed in this analysis represent a diversity of population sizes, forest areas and forestry and bioenergy production patterns. Both Canada and the United States have large forest sectors: most of the US forest landbase is privately owned, while most of the Canadian forest landbase is under public stewardship and forestry is conducted over a largely boreal, extensively managed and sparsely populated territory. The forest sector is an important part of both the Canadian and US economies, but Canada has a positive trade balance, while the United States has a negative trade balance. Sweden and Finland both have large forested areas and low population densities. Considering their boreal climate, Swedish and Finnish forests are very productive; forests are intensively managed and forest activities are highly mechanised. The forest sector is a vital part of Swedish and Finnish economies and both countries export large volumes of forest products. A significant share of their renewable energy is provided by forest biomass. By contrast, Norway has a forest landbase dominated by lightly managed natural forests and its forest sector is small, relative to size of its economy. Although a large fraction of the total energy supply is

TABLE 2.8 Forestry and Logging Employment and Productivity in 2011 (FAO, 2014)

Countries	Employment (1000 full-time employee equivalents)	Employment productivity (m^3 of roundwood per full-time employee equivalent)
Australia	11	2857
Belgium	3	1709
Canada	70	3153
Croatia	9	526
Denmark	4	517
Ireland	2	2031
Finland	23	1170
Germany	49	878
New Zealand	8	3733
Norway	7	1715
Sweden	20	2179
United States	160	3239

Note: Forestry and logging correspond to the ISIC Rev.4 Division 02 (forestry and logging) of the FAO classification of employment.

provided by renewable energy, bioenergy is of only minor importance. Croatia is dominated by productive public forests; its forest sector is small and less mechanised than that of the Nordic countries, but it is an important contributor to the national economy. Germany has very productive forests and a high population density. Though large and productive, the German forest sector is not as crucial to the economy as in Finland and Sweden and forest operations tend to be less mechanised than in the Nordic countries. Belgium and Denmark also have high population densities and small, privately owned, highly productive forests, dominated by plantations; these countries are large importers of forest products. However, Denmark is highly energy efficient and obtains a large share of its energy from renewable sources, particularly biomass. Australia is a large, sparsely populated country. Its forest landbase is mostly publicly owned. Although its forest sector is small, operations are highly mechanised, allowing the country to be a net exporter of forest products. In absolute terms, the forest sector in Ireland is also small, but quite dynamic. Renewable energy contributes only a small fraction of Ireland's total primary energy supply. New Zealand is a medium-sized forested country; wood production is concentrated in highly productive plantations and its large forest sector is focused on exports. Renewable energy represents an important share of the total primary energy supply, but the bioenergy sector itself is not large.

SUPPLY CHAIN DESCRIPTION

The following describes typical forest bioenergy supply chains in the studied countries, based on a mix of stakeholder surveys, expert knowledge and reference material from each country.

Australia

In Australia, bioenergy currently provides only a small fraction of total electricity supply; energy generation from wood occurs on a very small scale. Forest biomass-based energy supply chains are dedicated primarily towards heat production: wood chips for domestic use, wood pellets for export to Asia. The main feedstocks are harvesting residues (tree tops and branches that are by-products of industrial roundwood harvesting operations) in pine and eucalyptus plantations. At harvest, the average yield of roundwood is 450 m³/ha (average stand age of 30 years) in pine plantations and 200 m³/ha (average stand age of 15 years) in eucalyptus plantations. According to the Australian Forest Industries Research Centre's (FIRC) database, the amount of biomass available as harvesting residues can vary between 40 and 230 m³/ha in pine plantations and from 50 to 110 m³/ha in eucalyptus plantations.

Biomass feedstocks can also come from whole trees in low-quality stagnating eucalyptus plantations. Biomass harvesting operations are generally planned and executed independently of roundwood harvesting, although they could theoretically occur simultaneously on the same cutblock. There is no federal subsidy to provide financial assistance for growers of forest biomass. Furthermore, many growers, especially in Tasmania and in the Green Triangle, are concerned about the potential ecological consequences of removing high volumes of slash from pine plantations. According to local requirements, more than 50% of all harvesting residues must be left on-site. The low demand for forest bioenergy, lack of suitable biomass harvesting technologies and generally high operating costs are seen as important roadblocks that restrict the mobilisation of forest biomass supply chains. Also, the lack of good biomass inventory information is impeding development of business cases. Since 2011, however, the FIRC has been collecting and compiling data on the availability of harvesting residues, with the aim of building knowledge about the physical, ecological and techno-economical availability of forest biomass.

Belgium

In Belgium, 80% of the forest landbase is located in Wallonia, in the southern part of the country. Forests cover about one-third (550,000 ha) of Wallonia; about half of these are under private stewardship, while the other half are under public stewardship (regions, communes, etc.). Public forests are managed for high-quality wood production and environmental protection by the Wildlife and Forestry Department.

Overall, production of forest biomass for bioenergy is not a high priority for forest managers in Belgium. Nevertheless, mobilisation of modern bioenergy supply chains began in 2001, when the emergence of new products, such as wood chips and pellets, required that a more professional structure be put into place to meet the needs of the forest bioenergy sector (Marchal, 2010). Prior to this, the bioenergy supply chain was dominated by fuelwood used for domestic heating in rural areas, and wood processing residues used by industry to meet energy needs for heating, wood drying, etc.

At present, there are four typical biomass supply chains for bioenergy production in Belgium:

- fuelwood for domestic household heating;
- internal use of by-products (bark, sawdust, black liquor, etc.) in wood processing industries;
- wood chip production for community systems; and
- wood pellet production for domestic or industrial use.

Fuelwood production is based on the traditional model of wood harvesting for local use within a relatively short distance of forest woodlots. Although this type of activity is well developed, there are few available statistics because of its informal nature and the fact that wood is primarily harvested for self-consumption.

Wood chips are produced for community district heating systems (DHS) in rural areas. The main feedstock is low-quality roundwood that has no other commercial application. This roundwood comes from defective stems, coniferous trees from early thinnings, etc. Many companies are active in this sector and offer services such as roadside chipping, transport and storage.

As of 2015, there were seven wood pellet plants in Wallonia, with a total production capacity of 650,000 metric tonnes/year, although actual production in 2014 was an estimated 450,000 metric tonnes. The main feedstock for pellet production is sawdust from local sawmills. Wood pellets are used for domestic household heating, community heating and electricity production in industrial plants. However, imported wood pellets are also being used for electricity production in one large (80 MW) biomass plant.

The prospect of competition for wood to meet the needs of both the conventional wood products and the bioenergy industries is fuelling heated debate in Wallonia. Eventually, these discussions should lead to the development of an official strategy for forest bioenergy. Efforts are also being made to encourage small woodlot owners to adhere to principles of sustainable forest management.

Canada

For the most part, Canadian forest biomass supply chains are immature, with the notable exception of supply chains in which processing residues are used internally by the forest sector for heat and power production. Despite some niche applications, domestic use of bioenergy is very low. In fact, there is a lack of public awareness that biomass can be an efficient source of energy. Within the forest sector, there is a minimal understanding of fibre supply in the context of forest bioenergy and the larger bioeconomy, with some exceptions. As well, there is no coordinated national bioenergy policy or strategy. Nevertheless, some forest biomass supply chains do exist in Canada.

In terms of total volume, the production of wood pellets using processing residues from sawmills, as well as harvesting residues from clearcut operations, probably represents the largest forest biomass supply chain in Canada. Although a portion of these pellets are consumed domestically (eg for power production in the province of Ontario), most are shipped to Europe and, increasingly, to Asia. On a smaller scale, other supply chains are deployed to provide feedstock for district heating systems serving community, commercial and institutional clients. These supply chains use wood chips made from harvesting residues and from low-quality trees that have no merchantable value for other forest products. Most bioenergy feedstock supply chains in Canada are dependent on other forest products or forest practices. However, biomass harvesting operations are typically conducted as parallel processes, with different actors who are, at times, unwilling to collaborate, rather than being planned and coordinated in concert with roundwood harvesting operations and other silvicultural treatments.

The future availability of biomass feedstock is difficult to predict because feedstock sources are highly dependent on the markets for other forest products. Moreover, the sheer size of the Canadian forest landbase and the high variability found in its largely natural stands means that good quality forest inventory data are sparse. Natural disturbances (such as fires, insect outbreaks) that drive stand establishment across the Canadian forest landscape also produce large amounts of dead wood. A portion of this dead wood often remains unused by sawmills and pulpmills: the marginal quality of the fibre makes it difficult to process into conventional forest products and greatly reduces the profitability of mill operations. If bioenergy was added to the forest products basket, this marginal quality material could be utilised, as was successfully shown following the Mountain Pine Beetle epidemic in British Columbia, which has fuelled the development of a large pellet industry.

Croatia

Croatia has a small forest landbase, but benefits from a long history of sustainable forest management. Seventy eight per cent of all forests are publicly owned and are managed by Hrvatske šume (Croatian Forest Ltd.). The annual cut is set at 80% of the total annual increment, the equivalent of approximately 4.0 m^3/ha per year or 2.31 m^3/capita per year. However, the economic value of the forest sector remains low. In 2010, the forest sector contributed only 3.4 and 5.7% to value-added products and workplace demand in the manufacturing sector, respectively.

Targets for biomass use are described in its National Renewable Energy Action Plan (NREAP), but the legislation to support these targets is not complete. At present, biomass provides approximately 55% of all renewable energy in Croatia. Solid biomass comprises the largest share (91%) of biomass use, in large part because more than half of all Croatian households are heated with wood. Wood fuel remains the cheapest source of heating in Croatia.

More modern uses of biomass in Croatia include 7.69 MW_{el} in four installed solid biomass cogeneration plants, with more cogeneration plants (75.45 MW_{el}) and district heating plants pending. About 4% of the annual contracted biomass is directed towards heating applications, and the remainder will be used in cogeneration plants with mandatory minimum fuel efficiencies of 50%. Both existing and pending projects have contracts for 830,000 metric tonnes of solid biomass, close to Hrvatske šume's total annual cut. Nevertheless, these contracts are conditional on actual biomass use and could be terminated.

Modern forest biomass supply chains are still emerging in Croatia and, as a result, there are few reliable statistics. Given that it manages a large proportion of the forest landbase in Croatia, Hrvatske šume will be the main supplier of solid biomass. At present, it supplies wood chips, fuelwood and industrial roundwood for energy purposes (CHP, pellet production, household heating). Wood chips are produced directly in the forest, during forest management activities (with mobile chippers or adapted tractors with shredders), at the roadside (with a tractor or chipper) or at the client's yard. Waste timber and low-quality roundwood are typically provided at the roadside, or are delivered directly to the client. Fuelwood can be obtained at the roadside, with an extra charge for delivery. For example, the Lika Energo Eco CHP plant (0.95 MW_{el}) contracts roundwood and waste timber from Hrvatske šume and from local private foresters. Low-quality wood is designated for CHP and high-quality wood is designated for wood pellet production, which uses heat from the CHP plant. A similar supply chain is used by other pellet producers, but without CHP.

The second biomass supply chain in Croatia originates with the wood processing industry. For example, biomass for the largest (3 MW_{el}) CHP plant in Croatia originates with the production of parquet flooring. Processing residues are used to produce wood briquettes and to fuel the adjacent CHP. The flooring company also has a contract for wood chips from Hrvatske šume, in order to optimise the use of their own wood waste for export or for consumption. Most wood processing industries use their own woody residues to produce heat, briquettes or wood pellets. It is also common for these industries to offer wood waste as a salary supplement to workers.

Denmark

Biomass for bioenergy has been a part of Denmark's energy portfolio for centuries. However, it was the oil crisis of the early 1970s that prompted a major shift in the country's energy consumption, which at that time was dominated by imported fossil fuels. Biomass was incentivised and promoted as a renewable energy source and as an alternative to fossil fuels (Gregg et al., 2014).

The main market drivers are that there is no CO_2 tax on biomass and that the utility companies are obliged by a government decree to use straw and woody biomass in power production. Since 2001, a price subsidy of 15 øre/kWh (2 Euro cent/kWh) is in place. In the early biomass days in the 1980s and 1990s, there was heavy investment support (up to 25%) on new biomass plants, both for CHP and pure heat production plants. In addition, there has been heavy taxation of fossil energy since the 1970s (Stelte et al., 2015).

Production of bioenergy in Denmark has therefore increased steadily since 1990. Woody biomass and straw are the two main biomass sources. Until around 2000, they both came from domestic sources but, since 2000, woody biomass is being imported in considerable amounts. In 2013, 47.8 PJ of woody biomass was imported to Denmark, corresponding to 34% of the biomass utilised for energy (Stelte et al., 2015).

Denmark is the second largest importer of wood pellets to the European Union, after the United Kingdom (see chapter: Challenges and Opportunities for International Trade in Forest Biomass). Wood pellet imports correspond to more than 90% of the pellet consumption. Wood chips are also imported, covering about 25–30% of the demand. Wood pellets and wood chips are mainly imported from countries around the Baltic Sea, but also from North America and Africa. These feedstocks cover the increasing demand for private small-scale, medium-scale (district heating plants) and especially the large-scale (CHP) consumption. Fuelwood is also imported in smaller amounts from the Baltic area, by boat or truck (Bentsen and Stupak, 2013; Stelte et al., 2015).

Forestry in Denmark is a minor activity, compared to the scale of operations in other Nordic countries. There is no pulp industry and the harvested wood is mostly destined for sawlogs and bioenergy; about half of the harvested wood in both coniferous and broad leaved forests are used for energy. An important aspect of forest management, especially in states forests that constitute 25% of the forest area, is the protection of cultural and recreational values of the forest and 'close to nature' management, a concept that was introduced via the National Forest Program in 2002 (Skov-og Naturstyrelsen, 2002) and further defined by Larsen (2005). The use of native tree species and natural regeneration methods, permanent forest cover, development of diverse forest structures and individual tree management are key principles of this concept.

Forest bioenergy feedstock in coniferous stands is often procured from whole trees of first and second thinnings and logging residues of final; in broadleaved stands, branch residues are harvested as fuelwood, or, increasingly, chipped and removed for industrial energy production (Statistics-Denmark, 2013). Trees are typically felled during winter and chipped on site after drying at least one summer. When roundwood assortments are used for energy production, they are harvested by single grip harvester-forwarder system. Depending on the end-user, chipping is done at the roadside, at a terminal or at the end-use facility.

Many of the small forest owners are members of forest owner associations and have handed over their forest management to cooperatives who guide the forest management and hire contractors to conduct harvesting operations (Ramos, 2009).

Finland

Forest biomass supply chains are well established in Finland. Almost 9 Mm³ of residual materials are available for bioenergy production each year and include small diameter stems from early thinnings (4 Mm³), harvesting residues (3.5 Mm³) and stumps (1.1 Mm³). If pulpwood-sized stems were also used, as much as 20 Mm³ of woody biomass could be available for bioenergy production each year. Wood processing residues are already fully utilised for energy production; therefore, growth in the supply chain can mostly be achieved using biomass removed from the forest. That being said, the annual allowable cut is actually 30 Mm³ greater than the current harvesting rate. An increase in wood processing capacity to accommodate a greater rate of harvest would also significantly increase the availability of residues for bioenergy production.

The price of forest biomass-based energy in Finland is competitive with oil and gas but not, as of 2016, with coal. As a consequence, large CHP plants along the coast have started to use coal, rather than forest biomass, and some inland locations have started co-firing of forest biomass and coal. Nevertheless, the use of woody biomass for energy has increased steadily over the last several decades and, in 2013, provided 26% of total primary energy. In general, the price of woody fuels has been increasing slowly over the last few years, while demand has remained relatively stable.

Woody biomass for small heating plants in Finland is typically chipped at the roadside. There are nearly 1000 heating plants that rely on forest biomass for energy production in Finland, and most municipal centres use such heating plants. Larger CHP plants also tend to use wood chips but chipping is usually performed in large terminals, or at the CHP plant itself.

The machinery used to cut and forward materials for bioenergy production is the same machinery used to harvest and transport industrial roundwood. However, minor adjustments (eg to the cutting head of the harvester and the load space of the forwarder) are made to improve efficiency. Trucks are often used to transport wood chips to large power plants, but farm tractors can also transport chips to smaller plants. Terminals have been established to store and process wood for energy use and to ensure the security of supply year-round; this is particularly critical during wet or snowy conditions, when access to forest landings can be difficult and during periods of high demand (i.e. in the winter). In fact, the large variation in seasonal demand for woody biomass is a critical challenge for the forest biomass industry because harvesting machinery and trucks can be left to sit idle during periods of low demand.

To date, the domestic wood pellet market has been rather small, but that has started to change in recent years. In the past, most of the wood pellets produced in Finland were destined for export. Recently, however, Finnish wood pellets have been competing poorly with those produced in Russia and North America. As a consequence, Finnish pellet producers have been focusing on the domestic market, where large coal plants have begun substituting coal for locally produced wood pellets, with the aim of reducing CO_2 emissions.

The growing stock and, thus, the carbon stored in Finnish forests has been expanding rapidly (30–40 metric tonnes CO_2 equivalent/ha per year). This trend is expected to accelerate in the coming decades and, assuming that Finnish forests are managed sustainably, could mean that an even greater volume of raw materials and energy will be available to support the modern bioeconomy.

Germany

In Germany, several policies and financial incentives have been put in place to promote the use of renewable energy, including that from forest biomass. Although there are no direct subsidies for forest owners looking to enter the bioenergy market, there are subsidies to support CHP production and to encourage private homeowners to install wood pellet furnaces.

Most forest biomass supply chains in Germany are directed towards the production of bioenergy for domestic use. The raw materials entering the forest biomass supply chain include the residual tops and branches from partial cutting operations, conducted in natural/semi-natural mixedwood stands (with an average yield of 10.9 m^3/ha per year). The roundwood generated during these harvests is used for the production of conventional forest products while the harvesting residues are typically chipped for use in domestic community CHP systems. Another common forest biomass supply chain involves the traditional harvesting of fuelwood for domestic heating, which might also involve the recovery of roundwood for other uses (eg sawnwood). Industrial residues from sawmills, consisting of sawdust, wood chips slabs, etc., can be directed towards wood pellet production. Pellets are used domestically for small-scale heat production or for CHP systems, in private homes. The strong public opposition to nuclear power and fossil fuels has provided the momentum for deploying bioenergy systems. Moreover, the high price of energy and labour have led to a depression in the pulp and fibreboard industries, leading to an increase in the availability of forest biomass for bioenergy production. Forest bioenergy is also seen as a simple and inexpensive method of disposing of spruce bark beetle-infested trees. However, there are growing concerns about competition between the conventional forest product and forest bioenergy industries for fibre. Furthermore, the demand for wood is expected to increase over the next few decades and many scenarios predict a wood shortage in the near future. In order to facilitate the deployment of forest biomass and prevent a wood shortage, small-scale private forest owners will need to be mobilised and forest resources will need to be more efficiently utilised. This situation may also present an opportunity to develop integrated landscape management plans that would reduce competition within the forest sector.

Ireland

Ireland's forests are an important source of fibre for the timber industry and for energy generation. Irish forestry is highly productive, with an average yield class for Sitka spruce of 17 m³/ha per year (Farrelly et al., 2009). Sitka spruce (*Picea sitchensis*) is the most important and widely planted tree species in Irish forestry, occupying 52.3% of the total forest estate, or 327,000 ha. It is the dominant species planted during afforestation, accounting for around 60% of the national planting program since the 1970s. The mechanised cut-to-length (CTL) method predominates in Irish forest systems, accounting for approximately 95% of all timber harvesting (Jiroušek et al., 2007). Harvesting by the CTL system involves felling, delimbing and crosscutting by the harvester, followed by forwarding to the roadside. Secondary haulage is carried out by road or rail.

Currently, Ireland has a large number of young conifer plantations that are approaching the age of first thinning (Kofman, 2006). As a consequence, the net realisable volume from thinnings in Ireland is projected to increase from 1 million m³ over bark in 2011, to nearly 2 million m³ over bark in 2028 (Phillips, 2011) and, as such, will be an important source of forest biomass.

Biomass from harvesting residues, that is tops and branches, could also be a large source of biomass for energy production (Huang et al., 2013). In Ireland, forest residues have traditionally been left on clearcut sites. Occasionally, some of the larger residues are removed as fuelwood for domestic consumption, but this does not occur on a large scale. Interest in the collection and use of harvesting residues has increased in recent years, as demand for bioenergy has risen. Recent trials by Coillte, the state-owned forest company, estimate that up to 80 green tonnes/ha of harvesting residues could be recovered on suitable sites, depending on species, age, site type and the wood assortments harvested. It is estimated that raw material in the 'tip – 7 cm' category will increase from 48,000 m³ in 2011 to 61,000 m³ by 2020 (Phillips, 2011). However, this resource is only likely to be available on about 35% of harvested sites in Ireland, because of environmental constraints and restricted soil types (Murphy, 2014).

There is no stump harvesting currently carried out on a commercial scale in Ireland. However, research trials are being conducted by Coillte and Waterford Institute of Technology (WIT) on the feasibility and productivity of stump harvesting in Ireland. Stump harvesting results in increased intensification of forest management, when compared to conventional systems with above-ground biomass harvesting only. Benefits of stump harvesting include: increased production of wood energy resources, reduced CO_2 emissions when compared with fossil fuel consumption

(Eriksson and Gustavsson, 2008; Sathre and Gustavsson, 2011), improved site preparation and potential reduction of *Heterobasidion*, a fungus causing root disease (Walmsley and Godbold, 2010). However, the soil disturbance resulting from stump harvesting can also reduce the amount of carbon stored in forest soils, thus causing indirect CO_2 emissions (Melin et al., 2010; Repo et al., 2011), and can also deplete forest nutrient stocks (Walmsley and Godbold, 2010).

The two main biomass supply chains used in Ireland are:

- Standard short wood stems, produced during mechanical thinning: mechanical harvesting produces 3-m long delimbed stems and usually leaves a brush mat of branches and stem material less than 7 cm in diameter and 3 m length, on which the harvester and forwarder can drive to minimise soil disturbance. Chipping of the thinnings is carried out at the roadside, by tractor or truck-drawn machine. Wood chips are then fed by a crane fixed to the tractor or truck and transported directly to the plant, using walking floor trucks.
- Whole trees, produced during manual thinning: trees are felled without delimbing or crosscutting, and a terrain chipper is used to chip the whole tree at the stump. Such whole-tree harvests correspond to a row thinning, but with no selection between rows. Chipping is carried out by a Silvatec™ whole-tree terrain chipper, or equivalent machine. A chip forwarder loads the chips onto walking floor trucks, which deliver the wood chips to the power plant.

New Zealand

New Zealand has a relatively large forest industry, with almost 30 million m^3 of wood harvested annually, mostly from large-scale *Pinus radiata* D. Don plantations. With a strong silvicultural focus on producing sawlogs and limited outlets for lower-grade logs, large volumes of residues (estimated to be greater than 1.5 million Mm^3) and low-value logs (about 4 million m^3) could be used for bioenergy production. However, bioenergy currently contributes only a very small share of the total electricity supply (6%) and most of that is generated in-house at wood processing plants. Most of New Zealand's electricity supply is generated using hydro or geothermal. The low cost of electricity in New Zealand means that biomass is not competitive. As such, the government is not highly motivated to increase its production of bioenergy and does not provide any direct support.

Some biomass recovery from the forest does occur when larger volumes of good quality material can be readily accessed at a landing and delivered to a processing plant or mill in need of boiler fuelwood. However, only an estimated 10% of easily accessible residues are used in this way. Once these materials are collected, they are typically comminuted using a tub grinder, because of contamination with soil. The value of this biomass is expected to cover the cost of collection, transportation and comminution, but there is often not any additional value for the forest owner. Only in the most southern region of New Zealand, where winters are longer and colder, is there an established commercial market for wood fuel (chips or pellets). Pellets for the residential market are made almost exclusively from mill residues (ie sawdust or shavings) (Visser et al., 2009).

Norway

Norway is part of the European Economic Area (EEA), allowing the country to participate in the single European market. The Directive 2009/28/EC is EEA relevant, but no renewable targets have been set for Norway. Norway has large resources of oil and gas and is a large producer of primary energy. It aims to become carbon neutral by year 2030. The objective of the Norwegian energy strategy is to reach a bioenergy production of 100 PJ by 2020, almost doubling current production. This will result in a 72% share of renewable energy sources in final energy consumption by 2020, compared with the present 58% share (Rorstad and Trømborg, 2009). However, the abundance of hydro power and low electricity prices in Norway do not favour an increase in bioenergy production (Trømborg and Leistad, 2011).

The main sources of bioenergy are fuelwood [used for household heating (20 PJ)] and woody residues and waste [used for district heating (10 PJ) and by the forest industry (18 PJ)] (Rorstad and Trømborg, 2009; Trømborg and Leistad, 2011). Biomass from waste and from pulp processing residues is also used to produce approximately 1.8 PJ of electricity (Trømborg and Leistad, 2011). In fact, bioenergy is used to produce approximately 25% of the energy required by the wood processing industry. This energy is used for heating, as process energy, or for electricity production. Almost all of the processing residues, black liquor and waste wood generated by sawmills and other wood processing industries are used on-site to provide heat, process energy or electricity (Scarlat et al., 2011). Norway imports large quantities of pulpwood and wood chips from Russia and the Baltic States. A small share of this imported wood is also used for energy production, either as fuelwood or as bark, sawdust and black liquor, generated during processing for pulp and paper (Trømborg and Leistad, 2011).

Due to market conditions, biomass for energy production is not likely to be a driving force behind forest activities in Norway. Stump harvesting is controversial (Persson, 2012; Sverdrup-Thygeson and Framstad, 2007) and is not likely to be introduced in Norway in the near future. As well, small tree harvesting from pre-commercial thinning and early thinning activities is likely to be expensive. Therefore, any increase in the use of forest biomass for energy production will likely come from harvesting residues generated during conventional forest harvesting activities, such as clearcutting, seed tree and shelter wood harvests, and conventional thinning (Bergseng et al., 2013).

Sweden

Sweden, with more than 20 years of experience in the large-scale extraction of forest fuels for bioenergy production, has probably the most well-established forest biomass supply chains of all countries considered in this analysis. Annual supply volumes reach 6–7 Mm^3 and consist mainly of logging residues from final fellings (5–6 Mm^3) and small diameter trees from early thinnings (<1 Mm^3). To date, the use of stumps for energy has played only a minor role. Although processing residues from mills are completely used for energy production, there is still room for the industry to expand. The total annual volume of harvestable residual forest biomass may be well over 20 Mm^3; therefore, forest residues have the greatest growth potential, at least in the short term. Although short-rotation biomass production, mainly using willow, has been attempted, the area under cultivation remains small and has declined sharply in recent years in response to low prices for biomass and competition with agricultural uses.

The production of pulp and sawn goods have been increasing in Sweden over the last several years and the growth of forest bioenergy has, for the most part, capitalised on this growth. Bioenergy feedstocks from primary forest residues have proven competitive against oil and gas. CHP plants licenced for waste combustion have started to use large volumes of municipal waste, instead of forest biomass, resulting in a recent decline in the use of forest biomass-based fuels. Nevertheless, total forest biomass consumption continues to increase and, in 2013, produced 31% of total primary energy in Sweden. The price for wood fuels has been rising slowly but steadily over the last several years, as has the demand for wood-based fuel.

As in Finland, single-grip harvesters and forwarders are used to cut and extract residual forest biomass. Most of the wood chips used for heat and CHP plants in Sweden are produced using roadside chipping systems. Chip trucks transport the material to plants or to large terminals. The use of terminals for wood chip storage, sorting, comminution and trans-shipment of biomass is highly developed in Sweden and rail transport plays a major role in the long distance movement of forest chips.

The domestic wood pellet market in Sweden has been large, but has started to decline in recent years due to competition from wood pellets imported from the Baltic countries and from North America. The use of heat pumps also competes strongly with pellet-based residential heating systems. However, the main challenge to the economic viability of forest biomass-based energy is the entrance of solid municipal waste into the fuel market. To compete effectively in the future, forest biomass supply chains will need to become more efficient.

United States

There is no national, coordinated forest bioenergy policy in the United States. The supply chains and markets for conventional forest products are well established and, for the most part, the free market drives wood utilisation from privately owned and managed forests. However, there are numerous states with

Renewable Portfolio Standards (RPS) that set renewable energy targets for regulated electric utilities. In some states, these RPS created Renewable Energy Certificates (REC) that can provide incentives for biomass power production that meets certain criteria (eg air pollution control). In practice, however, REC are often met using other forms of renewable energy. On the other hand, the use of heating fuels, such as oil, propane and kerosene, remain largely unregulated and, as a result, there are fewer policy mechanisms at the state or federal level to deliver incentives to the bioenergy market. However, there is considerable interest in the development and utilisation of liquid biofuels.

A typical forest biomass supply chain, particularly in the Northeast and Southern United States, utilises residues (ie tree branches and tops) from commercial thinning or partial cutting of whole trees, in natural or semi-natural hardwood stands. At the time of harvest, the average standing volume of these stands can vary between 140 and 250 m^3/ha, while the amount of biomass available from the harvesting residues can vary between 25 and 37 m^3/ha. Harvesting residues stock-piled at the landing or roadside can be chipped and used as fuel by regional biomass power plants connected to the public power grid. Utilisation and processing of biomass for fuel is nearly always coordinated and integrated with conventional harvesting of merchantable roundwood.

Another typical supply chain produces wood pellets for international trade, mainly to Europe. Most of the biomass used for wood pellet production originates in pine plantations in the Southern and Southeastern (SE) United States, which have average standing volumes of 173–247 m^3/ha. The supply chain is based on the use of material generated during commercial thinning and harvesting of whole trees. High-quality logs are used for conventional forest products and low-quality logs are used for wood pellets. The wood pellet market is gradually replacing the pulp and paper market in the SE United States and the supply chain mimics the traditional pulpwood model.

In general, there are no direct regulations or quotas controlling the rate of forest biomass utilisation in the United States; instead, forest biomass extraction is driven largely by market economics. As such, demand from Europe for wood pellets is the main driver for the deployment of forest biomass supply chains in the SE United States. However, the development of sustainability standards in Europe (such as the standards aimed at reducing carbon emissions) is considered a potential risk to the further mobilisation of supply chains in the region. As well, the United States has a long history of public debate and dialogue regarding the sustainable use of its forest resources and has implemented a robust suite of policies, monitoring systems and reporting activities over the last century, aimed at ensuring sustainable forest management. The establishment of the USDA Forest Service and associated public lands management, in 1903, has fostered decades of forest research, an annualised Forest Inventory system, federal assistance programs to private farm and forest landowners and a suite of federal laws and regulations related to the protection of air, water and endangered species. These federal investments are supported by equally substantial state and local government investments, including development of best management practices (BMPs), state forest action plans and, in some cases, harvest or logger certifications, permitting and/or training programs. Some states, notably in the Northeast, have tax policies that encourage private forest owners to manage their forests actively and harvest timber periodically. Public policies, such as 'current use' property tax relief programs, provide benefits to the entire forest products industry, including bioenergy, and help counteract forest parcelisation and fragmentation.

SUMMARY AND CONCLUSIONS

It is often assumed that mobilisation of supply chains relies heavily on economic factors that determine the ease with which biomass can be procured, processed and delivered to markets. However, as seen in this chapter, many other interacting aspects of supply and demand can affect the level of forest biomass mobilisation in a region. Countries of the boreal and temperate biomes display widely different conditions in terms of their land and forest use, industrial development and energy policy. There is also considerable variability in terms of the organisation and maturity of their forest biomass markets and the share of the forest sector captured by bioenergy production. As well, there is significant variation in the degree to which bioenergy is integrated within the basket of wood products (along with conventional products, such as sawnwood and paper) in strategic and operational decision-making. These differences represent a challenge when it comes to making recommendations that are meaningful and applicable across an array of conditions. Nevertheless, as suggested by Jessup and Walkiewicz (2013), they also represent opportunities to form cross-regional and international synergies among stakeholders and markets with different but complementary characteristics, to transfer technologies and to develop niche applications or efficiencies that could result in large collective gains to global biomass mobilisation.

REFERENCES

Bentsen, N., Stupak, I., 2013. Imported Wood Fuels—A Regionalized Review of Potential Sourcing and Sustainability Challenges. University of Copenhagen, Frederiksberg.

Bergseng, E., Eid, T., Løken, Ø., Astrup, R., 2013. Harvest residue potential in Norway—a bio-economic model appraisal. Scand. J. Forest Res. 28 (5), 470–480.

Eriksson, L.N., Gustavsson, L., 2008. Biofuels from stumps and small roundwood—costs and CO_2 benefits. Biomass Bioenergy 32 (10), 897–902.

FAO, 2010. Global Forest Resources Assessment 2010 Main Report. FAO Forestry Paper 163. Food and Agriculture Organization of the United Nations, Rome, Italy, Available from: http://www.fao.org/docrep/013/i1757e/i1757e.pdf.

FAO, 2014. Contribution of the Forestry Sector to National Economies, 1990–2011. Forest Finance Working Paper FSFM/ACC/09. Food and Agriculture Organization of the United Nations, Rome, Italy. Available from: http://www.fao.org/3/a-i4248e.pdf

FAO, 2015. FAOSTAT. Available from: http://faostat3.fao.org/home/E

Farrelly, N., Ní Dhubháin, Á., Nieuwenhuis, M., Grant, J., 2009. The distribution and productivity of Sitka spruce (*Picea sitchensis*) in Ireland in relation to site, soil and climatic factors. Irish Forestry 64, 51–73.

Gregg, J.S., Bolwig, S., Solér, O., Vejlgaard, L., Gundersen, S.H., Grohnheit, P.E., 2014. Experiences With Biomass in Denmark. Technical University of Denmark, Roskilde, Denmark.

Huang, Y., McIlveen-Wright, D.R., Rezvani, S., Huang, M.J., Wang, Y.D., Roskilly, A.P., Hewitt, N.J., 2013. Comparative techno-economic analysis of biomass fuelled combined heat and power for commercial buildings. Appl. Energy 112 (0), 518–525.

IEA, 2015. World Energy Statistics and Balances. Available from: http://dx.doi.org/10.1787/data-00510-en

Jessup, E.L., Walkiewicz, J., 2013. Adapted forestry practices for improved biomass recovery. D3.2. INFRES—Innovative and Effective Technology and Logistics for Forest Residual Biomass Supply in the EU. Available from: http://www.infres.eu/en/results/

Jiroušek, R., Klvac, R., Skoupý, A., 2007. Productivity and costs of the mechanised cut-to-length wood harvesting system in clear-felling operations. J. Forest Sci. 53 (10), 476–482.

Kofman, P.D., 2006. Harvesting Wood for Energy From Early First Thinnings. Available from: http://www.coford.ie/publications/cofordconnects/

Larsen, J.B., 2005. Natural forest management (in Danish: Naturnær skovdrift). Dansk Skovbrugs Tidsskrift 90. Dansk Skovforening (Danish Forestry Association), Copenhagen, Denmark. 400 p. Available from: http://www2.sns.dk/udgivelser/2005/naturaerskovdrift_jbolarsen.pdf

Marchal, D., 2010. Le bois-énergie, aspects historiques et tendances actuelles. Forêt Wallonne 104, 4–13.

Melin, Y., Petersson, H., Egnell, G., 2010. Assessing carbon balance trade-offs between bioenergy and carbon sequestration of stumps at varying time scales and harvest intensities. Forest Ecol. Manag. 260 (4), 536–542.

Murphy, F., Devlin, G., McDonnell, K., 2014. Forest biomass supply chains in Ireland: a life cycle assessment of GHG emissions and primary energy balances. Appl. Energy 116, 1–8.

Persson, T., 2012. Tree stumps for bioenergy—harvesting techniques and environmental consequences. Scand. J. Forest Res. 27 (8), 705–708.

Phillips, H., 2011. All Ireland Roundwood Production Forecast 2011–2028. COFORD, Department of Agriculture, Fisheries and Food, Dublin, Available from: http://www.coford.ie/media/coford/content/publications/projectreports/forecast_31Jan11.pdf.

Ramos, J., 2009. Sustainability and operational aspects of forest biomass harvesting for energy in Scandinavia. J. W. Gottstein Memorial Trust Fund. Available from: http://gottstein.nuttify.com/wp-content/blogs.dir/651/files/2012/09/jramos.pdf

Repo, A., Tuomi, M., Liski, J., 2011. Indirect carbon dioxide emissions from producing bioenergy from forest harvest residues. GCB Bioenergy 3 (2), 107–115.

Rorstad, P., Trømborg, E., 2009. Bioenergy markets and frame conditions in Norway. Workshop: Biomass resources and bioenergy in Norway and other Nordic countries, Oslo, Norway, September 23–25, 2009. Available from: http://iet.jrc.ec.europa.eu/remea/events/biomass-resources-and-bioenergy-norway-and-other-nordic-countries

Sathre, R., Gustavsson, L., 2011. Time-dependent climate benefits of using forest residues to substitute fossil fuels. Biomass Bioenergy 35 (7), 2506–2516.

Scarlat, N., Dallemand, J.-F., Skjelhaugen, O.J., Asplund, D., Nesheim, L., 2011. An overview of the biomass resource potential of Norway for bioenergy use. Renew. Sustain. Energy Rev. 15 (7), 3388–3398.

Skov-ogNaturstyrelsen, 2002. Danmarks nationale skovprogram (in Danish: Denmarks national forest program). 82 p. Available from: http://www2.sns.dk/udgivelser/2002/87-7279-452-6/danmarks_nationale_skovprogram.pdf

Statistics-Denmark, 2013. FOREST 6: Felling in Forests and Plantation in Denmark by Region, Species and Area. Available from: http://www.statistikbanken.dk/statbank5a/default.asp?w=2021

Stelte, W., Hinge, J., Dahl, J., 2015. Country Report 2014 for Denmark: IEA Bioenergy Task 40. Available from: http://www.bioenergytrade.org/downloads/iea-task-40-country-report-2014-denmark.pdf

Sverdrup-Thygeson, A., Framstad, E., 2007. Bioenergy Measures and Effects on Biodiversity (in Norwegian With English summary: Bioenergitiltak og effekter på biomangfold). Norsk institutt for naturforskning (NINA), Oslo, Available from: http://www.nina.no/archive/nina/PppBasePdf/rapport%5C2007%5C311.pdf.

Trømborg, E., Leistad, Ø., 2011. Country Report 2011 for Norway: IEA Bioenergy Task 40. Available from: http://www.bioenergytrade.org/downloads/iea-task-40-country-report-2011-norway.pdf

Visser, R., Spinelli, R., Stampfer, K., 2009. Integrating biomass recovery operations into commercial timber harvesting: the New Zealand situation. Council on Forest Engineering Conference Proceedings: Environmentally Sound Forest Operations, Citeseer.

Walmsley, J.D., Godbold, D.L., 2010. Stump harvesting for bioenergy—a review of the environmental impacts. Forestry 83 (1), 17–38.

Chapter 3

Quantifying Forest Biomass Mobilisation Potential in the Boreal and Temperate Biomes

David Paré*, Evelyne Thiffault**, Guillaume Cyr*, Luc Guindon*
*Canadian Forest Service, Natural Resources Canada, Quebec City, Canada; **Department of Wood and Forest Sciences and Research Centre on Renewable Materials, Laval University, Quebec City, Canada

Highlights

- Indicators based on ecosystem net primary production, roundwood production and solid biofuel production can be used to estimate the potential availability of forest bioenergy feedstocks for a given country.
- Approximately 13.5 EJ of additional energy could be produced, if studied countries achieved the same level of forest management and biomass recovery as the best performing countries. A similar increase in the intensity of roundwood production and recovery of forestry by-products/residues in Russia could yield an additional 10.3 EJ.
- The feasibility of such mobilisation efforts depends on both national and regional constraints which can include environmental regulations, socio-economic considerations and technological limitations.

INTRODUCTION

Recent estimates of the global potential for biomass deployment have varied widely (Bentsen and Felby, 2012; Berndes et al., 2003; Chum et al., 2011; Creutzig et al., 2014; Hoogwijk et al. 2003; Slade et al., 2014), in part because each study has relied on a different suite of assumptions and drivers. The supply of woody biomass can be subject to a range of theoretical, technical, economic and ecological limitations (Smeets et al., 2007). For example, theoretical woody biomass potential is constrained by natural and climatic parameters, and is then netted down by different factors to further refine estimates (Verbruggen et al., 2011) (Box 3.1).

BOX 3.1 Definitions of Forest Biomass Potential (Adapted from Smeets et al., 2007)

- Theoretical potential: the maximum wood production potential of forests, as determined by prevailing conditions for biological growth.
- Technical potential: the theoretical potential, constrained by limitations imposed by technical barriers.
- Economic potential: the technical potential under specific market conditions, for example below a maximum production cost.
- Environmental/ecological–economical potential: the economic potential, constrained by limits on harvest residue removal (to protect soil fertility) or establishment of ecological reserves (to protect biodiversity).

As mentioned in chapter: Introduction, sustainable forest biomass production can include: (1) the current rate of production of roundwood for conventional forest products, such as sawnwood, paper and fibreboard; (2) additional roundwood that could be harvested within the sustainable harvest limit or annual allowable cut (AAC: the annual amount of timber that can be harvested on a sustainable basis within a defined forest area); (3) primary forest residues that are by-products of silvicultural and harvesting operations, such as tree tops, branches, small trees from early thinnings; (4) secondary forest residues from industrial wood processing; and (5) tertiary residues, such as non-recyclable paper and demolition wood.

The potential availability of these five streams is interdependent and is determined by forest ecosystem productivity, the volume of roundwood harvested for conventional forest products and the product portfolio (eg the ratio of mechanically/chemically-produced pulp) and technical infrastructure (eg the equipment for process heat production). This chapter focuses on the first four streams; the use of tertiary residues for bioenergy production is not discussed further.

The purpose of the analyses presented in this chapter is first to provide estimates of the levels of biomass and bioenergy production currently achieved in several countries of the temperate and boreal biomes; second, it aims to inform on the potentials of forest biomass feedstocks that could be mobilised for bioenergy production in a profitable and sustainable manner. These estimates are calculated to facilitate first-order comparisons between countries and to show how mobilisation potentials are determined by the size of the forestry and bioenergy sectors, relative to the theoretical productive capacity of forest ecosystems in a given country. The group of countries in the boreal and temperate biomes that are included represent contrasting vegetation conditions, forestry practices and bioenergy policies (also see chapter: Comparison of Forest Biomass Supply Chains From the Boreal and Temperate Biomes): Australia, Belgium, Canada, Croatia, Denmark, Finland, Germany, Ireland, New Zealand, Norway, Sweden and the United States.

METHODOLOGY

For the purpose of this analysis, we developed a methodology that uses data routinely compiled by international agencies. Two indicators of forest biomass feedstock potential were developed for each country:

- the ratio of roundwood production to forest ecosystem net primary production (NPP), and
- the ratio of forest bioenergy production to roundwood production.

Roundwood-to-NPP Ratio

The ratio of roundwood production to forest ecosystem NPP for each country is unit-less, because both the numerator and denominator are expressed as (metric) tonnes of carbon (C)/year. NPP represents the net amount of C assimilated by vegetation over a given period of time; it determines the amount of energy that could be transferred from plants to other trophic levels and, thus, can be used to represent theoretical maximum forest production. Consequently, the roundwood production-to-NPP ratio provides an indicator of the intensity of forest management within a country, while controlling for limits to forest productivity, such as climate, soil and forest area. Similar approaches have been used to study human appropriation of biological production (Wright, 1990; Haberl et al., 2007; Imhoff et al., 2004; Krausmann et al., 2008; Rojstaczer et al., 2001; Vitousek et al., 1986) and to quantify potential bioenergy supply (DeLucia et al., 2014; Field et al., 2008).

Forest NPP

Global terrestrial NPP has remained relatively unchanged for at least 30 years (Ito, 2011; Running, 2012). On an areal basis, NPP represents approximately the upper limit of ecosystem productivity. NPP data would not be as useful for estimating crop production potential in agricultural systems (including intensive energy crops) because irrigation, fertilisation and other agronomic inputs can strongly influence yield and cropping intensity.

National estimates of forest NPP were derived using Moderate-Resolution Imaging Spectroradiometer (MODIS) satellite information, with a pixel resolution of 1 km (Heinsch et al. 2003), overlaid with the forest cover map generated by Hansen et al. (2013), with a pixel resolution of 30 m. Using this information, a raster grid (0.00833 deg. ~930 m) was produced, with estimates of the proportion of forest cover for each 1 km cell. Only cells with more than 80% forest cover were used for further analysis, to avoid over- or under-estimation of forest NPP. For all 12 countries discussed in this analysis, the number of cells with greater than 80% forest cover was sufficient to generate meaningful estimates of NPP. The median NPP value (in tonnes of C/ha per year) for each country was then multiplied by the surface area (in hectares) covered by forests (FAO, 2014) to obtain an estimate of total annual NPP for each country (Table 3.1).

TABLE 3.1 Estimated NPP for Forest Ecosystems

Countries	Forest land area (data for 2012) (ha)	Median NPP (tonnes of C/ha per year)	Total NPP (tonnes of C/year)
Australia	147,452,000	15.9	2,342,569,924
Belgium	679,880	8.0	5,472,354
Canada	310,134,000	5.2	1,626,652,830
Croatia	1,923,000	6.0	11,532,231
Denmark	548,000	7.2	3,946,148
Finland	22,157,000	4.7	105,002,023
Germany	11,076,000	7.5	82,715,568
Ireland	756,600	9.0	6,834,368
New Zealand	8,252,200	16.2	133,339,048
Norway	10,217,800	5.1	51,967,731
Sweden	28,203,000	5.7	159,995.619
United States	304,787,600	5.9	1,794,589,389

Roundwood Production

National roundwood production data were obtained from the FAO's national database of forest products (FAO, 2014), using averages calculated for the years 2002–2013. The FAO's definition of roundwood includes all roundwood felled or otherwise harvested and removed (eg wood used for fuel, charcoal and industrial processing). Roundwood production data were converted from m^3/year (excluding bark) to tonnes of C/year, using a default factor of 0.5 tonne of dry matter per m^3 of fresh volume, and assuming that 50% per unit mass of dry matter is C (Table 3.2).

Bioenergy-to-Roundwood Ratio

The ratio of bioenergy production to roundwood production is also unit-less, because both the numerator and denominator are expressed as tonnes of C/year. This index can be used to indicate the quantity of bioenergy feedstock resources that is mobilised relative to the quantity of roundwood produced for each country.

Forest Bioenergy Production

Primary solid biofuel is defined as any plant matter used directly as fuel, or converted into other forms before combustion; this can include materials generated by industrial processes, or provided directly by forestry and agriculture (eg. fuelwood, wood chips, bark, sawdust, shavings, black liquor and other solid biofuels). Average values of domestic primary solid biofuel (excluding

TABLE 3.2 Average Annual Roundwood Production (2002–2013) in the Studied Countries

Countries	m³/year	Tonnes of C/year
Australia	30,896,917	7,724,229
Belgium	4,871,785	1,217,946
Canada	164,961,846	41,240,461
Croatia	4,379,000	1,094,750
Denmark	2,346,102	586,525
Finland	52,087,018	13,021,755
Germany	55,297,980	13,824,495
Ireland	2,597,881	649,470
New Zealand	22,533,417	5,633,354
Norway	9,814,067	2,453,517
Sweden	71,699,917	17,924,979
United States	395,818,035	98,509

Source: FAO, 2014

charcoal) production data for 2002–2013, compiled for each country by the International Energy Agency (IEA, 2014) and expressed in terajoules per year (TJ/year), were used as a proxy for forest bioenergy production. Values were converted from TJ to tonnes of C/year, using the default emission factor of 112 tonnes of CO_2/TJ for wood and wood waste, provided by the Intergovernmental Panel on Climate Change in its Guidelines for National Greenhouse Gas Inventories (IPCC, 2006), and the molar mass of C (12 g/mol), relative to the total molar mass of CO_2 (44 g/mol) (Table 3.3).

Primary solid biofuels were assumed to originate exclusively from forests. This is roughly valid for most of the countries from the boreal and temperate biomes. However, significant quantities of agricultural residues are used as solid fuels in Denmark. To address this anomaly, the proportion of energy produced from straw in Denmark [approximately 23% (DEA, 2014)] was subtracted from estimates of solid biofuel production.

ANALYSIS

Table 3.1 presents the median and total NPP for each of the studied countries. Median NPP varies from 4.7 tonnes of C/ha per year for Finland to 16.2 tonnes of C/ha per year for Australia. Fig. 3.1 shows the distribution of NPP values for each country, using Kernel density estimates. Many countries show a near-normal (ie Gaussian) distribution of NPP. The distribution curves for

TABLE 3.3 Average Annual Bioenergy Production From Forests (2002–2013) in the Studied Countries

Countries	TJ/year	Tonnes of C/year
Australia	185,756	5,674,006
Belgium	32,466	991,679
Canada	483,304	14,762,749
Croatia	15,555	475,135
Denmark	44,324	1,353,890
Finland	305,549	9,333,136
Germany	376,113	11,488,544
Ireland	7,023	214,533
New Zealand	47,949	1,464,638
Norway	46,107	1,408,351
Sweden	347,525	10,615,303
United States	2,061,437	62,967,516

Source: IEA, 2014

Australia, Canada, Ireland, New Zealand and the United States display a broad base (platykurtic distribution), indicating that forest ecosystem productivity encompasses a wider range of conditions than in other countries (eg Belgium and Finland). Denmark and Ireland display a binomial distribution in NPP, which may reflect differences in forest productivity related to (1) planted

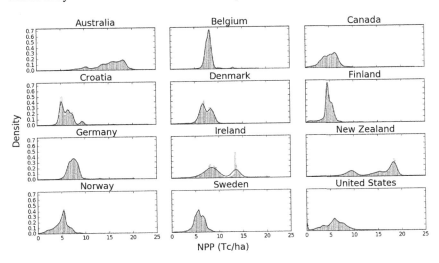

FIGURE 3.1   Kernel density distribution of forest NPP.

versus naturally regenerated forests; (2) forest ownership (see Fig. 2.2 and 2.3 in Chapter: Comparison of Forest Biomass Supply Chains From the Boreal and Temperate Biomes); and (3) type of forest (eg commercial plantations, native woodlands, urban forests). Australia and New Zealand show skewed distributions, which may reflect widely varying climatic conditions and forest types within these countries; forests in large parts of Australia and New Zealand fall within the Mediterranean and tropical biomes (see Fig. 1.4 in chapter: Introduction) and, as such, have much different rates of NPP than those in temperate biomes. Thus, Median NPP is not a perfect estimate of NPP for some countries, particularly when NPP data have a non-normal distribution, but can be considered a suitable approximation.

Table 3.1 shows that median NPP values for boreal countries (eg Canada, Finland, Norway and Sweden) are relatively similar, and are lower than those for temperate countries (eg Denmark, Belgium, Germany, Ireland). Median NPP for the United States falls between those from boreal and temperate countries; this is not unexpected, given that forests in the United States span both boreal and temperate biomes (see Fig. 1.4 in chapter: Introduction). Median NPP values for Australia and New Zealand are greater than for any of the other countries discussed here, because of their warm climates and highly productive tree species.

The ratio of roundwood production-to-NPP (Table 3.4) gives an indication of forest management intensity within each country. Calculated values of roundwood production-to-NPP were greatest for Belgium and Germany, for most Nordic countries (Denmark, Finland and Sweden) and for Ireland: a relatively

TABLE 3.4 Roundwood-to-NPP and Bioenergy-to-Roundwood Ratios

Countries	Roundwood-to-NPP (%)	Bioenergy-to-roundwood (%)
Australia	0.3	74
Belgium	22.3	81
Canada	2.5	36
Croatia	9.5	43
Denmark	14.9	231
Finland	12.4	72
Germany	16.7	83
Ireland	9.5	33
New Zealand	4.2	26
Norway	4.7	57
Sweden	11.2	59
United States	5.5	64

large share of forest NPP ($\geq 10\%$) is harvested for wood products each year. These results are consistent with the conclusions of the EU Biomass Futures project (Elbersen et al., 2012). By contrast, calculated values of roundwood production-to-NPP for Australia, Canada, New Zealand and Norway indicate that $\leq 5\%$ of forest NPP is harvested for wood products each year.

In part, differences in the roundwood production-to-NPP ratio among the studied countries reflect forest management intensity, for example the proportion of intensively managed plantations, relative to less intensively managed planted or naturally regenerated forests, or the proportion of the AAC that is actually harvested. In New Zealand, for example most of the roundwood is produced from fast-growing *Pinus radiata* plantations; since 2000, logging in native forests has only been conducted on private land and only if it can be shown to meet sustainability criteria. In Canada, natural disturbances (eg from insects or wildfire) affect large areas of forest each year (NRCAN, 2014), capturing a significant share of forest ecosystem NPP.

The ratio of bioenergy-to-roundwood production (Table 3.4) also varies widely among studied countries. Bioenergy production from primary (ie by-products of silvicultural and harvesting operations) and secondary (ie by-products of wood processing operations) forest residues are generally assumed to represent between 30 and 60% of the quantity of wood used for merchantable roundwood production (DOE, 2011; Lauri et al., 2014; Smeets et al., 2007). Indeed, most of the calculated bioenergy-to-roundwood production ratios listed in Table 3.4 fall within or close to this range. Countries that import large quantities of roundwood, relative to domestic production, such as Denmark and Belgium (Table 3.5), exhibit very high ratios of bioenergy-to-roundwood production; this is likely explained by the fact that secondary residues produced during the processing of imported roundwood can be important sources of feedstock for bioenergy. This indirect trade in biofuels can be significant (Heinimö, 2008): in 2004, approximately 22% of all wood-based energy production in Finland was derived from imported wood. Conversely, countries that export large quantities of roundwood exhibit low ratios of bioenergy-to-roundwood production (eg New Zealand; Table 3.5). Given the substantial flow of wood imports and exports among countries, the ratios of bioenergy-to-roundwood production reported here must therefore be interpreted with caution.

The type and intensity of forest management (eg the frequency of thinning; the frequency of primary residue recovery after thinning and final felling; the proportion of residues removed during such operations) are important drivers of the bioenergy-to-roundwood production ratio for each country. For example, the operational rate of primary residue recovery during clearcut harvesting of mature stands (ie the proportion of harvesting residues recovered from a cutblock) is approximately 50% in Canada and 70% in Nordic countries (Thiffault et al., 2014). In Nordic countries, several decades of functioning markets for forest bioenergy feedstocks have fostered the development of equipment and know-how aimed specifically at recovering residues from

TABLE 3.5 Domestic Production, Export and Import of Roundwood for Studied Countries

Countries	Domestic production (m³)	Export (m³)	Export of domestic production (%)	Import (m³)	Import of domestic production (%)
Australia	28,504,000	1,395,691	4.90	1,648	0.01
Belgium	5,128,001	1,077,620	21.01	4,940,920	96.35
Canada	148,183,000	6,137,000	4.14	4,588,000	3.10
Croatia	5,714,000	972,000	17.01	12,000	0.21
Denmark	2,446,312	599,051	24.49	649,519	26.55
Finland	52,309,505	696,886	1.33	5,552,685	10.62
Germany	52,338,132	3,511,647	6.71	6,965,568	13.31
Ireland	2,580419	201,880	7.82	220,424	8.54
New Zealand	27,415,000	13,758,897	50.19	3,601	0.01
Norway	10,572,145	1,627,999	15.40	1,056,002	9.99
Sweden	69,499,000	819,813	1.18	7,334,000	10.55
United States	387,512,000	14,711,300	3.80	1,279,026	0.33

Data for 2012.
Source: FAO, 2014

cutblocks. As well, intensively managed planted forests typically produce greater quantities of primary residues (eg during pruning and pre-commercial thinning) than naturally regenerated, less intensively managed forests (Hakkila and Parikka, 2002).

The structure of the forest industry can also have a strong influence on the ratio of bioenergy-to-roundwood production through its influence on the use of secondary forest residues. In eastern Canada, for example a large proportion of residues from sawlog wood processing is utilised as feedstock for the fibreboard and pulp industries.

SUMMARY AND CONCLUSIONS

The quantification method and results presented earlier reflect how mobilisation of forest biomass can be investigated at two levels:

- The intensification of forest management activities, in which forestry would appropriate a larger share of forest ecosystem net primary productivity; and

- The intensification of biomass recovery from silvicultural, harvesting and wood processing operations, in which bioenergy would appropriate a larger share of forestry by-products/residues.

As will be discussed in subsequent chapters, forest bioenergy is likely to remain a product that is strongly related to the management of the forest for conventional wood products. The results show that there is a significant potential to increase biomass feedstock supply for bioenergy production in many countries of the boreal and temperate biomes, even without an increase of roundwood harvesting. Notably, improved guidance and training of forest workers, design and access to better adapted harvesting and wood processing machinery and technology that allow for optimal recovery of by-products might increase the ratio of bioenergy-to-roundwood production and, therefore, increase mobilisation of forest bioenergy.

As noted, uncertainties due to export–import flows of wood and forest products prevents firm conclusions on achievable bioenergy-to-roundwood production ratio in different countries. Nevertheless, for the purpose of illustrating magnitudes, Table 3.6 shows the level of bioenergy supply in the studied

TABLE 3.6 Gain in Forest Bioenergy Production Relative to Current Production if the Ratio of Bioenergy-to-Roundwood Production Reached 83% (and Assuming no Change in the Ratio of Roundwood Production-to-NPP)

Countries	Projected bioenergy production (tonnes of C/year)	Gain (tonnes of C/year)	Gain (TJ^{-1})
Australia	6,418,834	744,829	24,384
Belgium	1,012,113	20,434	669
Canada	34,270,823	19,508,075	638,657
Croatia	909,737	434,603	14,228
Denmark	487,403	—	—
Finland	10,821,078	1,487,942	48,712
Germany	11,488,155	—	—
Ireland	539,710	325,177	10,646
New Zealand	4,681,317	3,216,680	105,308
Norway	2,038,872	630,522	20,642
Sweden	14,895,658	4,280,355	140,131
United States	82,231,197	19,263,680	630,656
Total	169,794,898	49,912,296	1,634,034

countries, if all reached at least the same ratio of bioenergy-to-roundwood production as Germany (ie 83%; Table 3.4). As can be seen, the total bioenergy supply would increase by roughly 1.6 EJ, or almost one-third.

The results of this analysis also suggest that forests in several countries are already highly managed. High ratios of roundwood production-to-NPP are typical of countries that employ more intensive silvicultural practices with frequent interventions, allowing a greater proportion of ecosystem productivity to be diverted into wood products. Increasing the roundwood production-to-NPP ratio, while maintaining the same ratio of bioenergy-to-roundwood production, could generate a significant increase in bioenergy supply. Table 3.7 shows that the total bioenergy supply could increase by approximately 8.6 EJ, if all of the countries included in this analysis attained a roundwood production-to-NPP ratio of 10% (equivalent to the current average ratio of the studied countries). Furthermore, the total bioenergy supply could increase by approximately 13.6 EJ, if all of the countries attained both a roundwood production-to-NPP ratio of 10% and a bioenergy-to-roundwood production ratio of 83%.

TABLE 3.7 Gain in Forest Bioenergy Production Relative to Current Production if the Ratio of Roundwood Production-to-NPP Reached 10% (and Assuming no Change in the Ratio of Bioenergy-to-Roundwood Production)

Countries	Projected roundwood production (tonnes of C/year)	Gain in roundwood (tonnes of C/year)	Gain in bioenergy (tonnes of C/year)	Gain in bioenergy (TJ/year)
Australia	234,256,992	226,532,763	166,404,717	5,447,773
Belgium	547,235	—	—	—
Canada	162,665,283	121,424,822	43,466,151	1,422,999
Croatia	1,153,223	58,473	25,378	831
Denmark	394,615	—	—	—
Finland	10,500,202	—	—	—
Germany	8,271,557	—	—	—
Ireland	683,437	33,967	11,220	367
New Zealand	13,333,905	7,700,551	2,002,096	65,545
Norway	5,196,773	2,743,256	1,574,665	51,552
Sweden	15,999,562	—	—	—
United States	179,458,939	80,504,430	51,227,216	1,677,081
Total	632,461,723	438,998,262	264,711,443	8,666,148

Large forest countries from the temperate and boreal biomes were not included in this analysis. For example 20% of the world's forest estates are located in Russia (FAO, 2010). Russia's wood pellet industry expanded sharply between 2003 and 2013 and, as a consequence, Russia has become an important wood pellet exporter to Nordic and Baltic countries (Proskurina et al., 2015). Median forest ecosystem NPP in Russia is 4.6 tonnes of C/ha per year and Russia's forested area covers 809.2 million ha. Between 2002 and 2013, average roundwood production in Russia was 189,861,583 m^3/year, while solid biofuel production reached 139,175 TJ/year. Using the same methods described earlier, estimates of roundwood production-to-NPP and bioenergy-to-roundwood production for Russia are 1.2 and 9.4%, respectively. Illegal logging is estimated to account for up to 10% of total roundwood production in Russia (Newell and Simeone, 2014). Nevertheless, these indices suggest that there is a massive potential for enhancing mobilisation of forest bioenergy feedstocks: an additional 10.3 EJ of bioenergy could be produced each year by increasing roundwood production to 10% of NPP and raising the ratio of bioenergy-to-roundwood production to 83%. However, there remain several barriers to the expansion of the Russian forest bioenergy industry. The market structure of the Russian forest industry is dominated by a small number of players; this has resulted in reduced competition and inflated consumer prices. Furthermore, there remains inadequate infrastructure and a lack of foreign investment (Proskurina et al., 2015). In order to close the gap between current and projected forest bioenergy production levels, several barriers at the local and international level will need to be overcome. In fact, these challenges are shared, to a greater or lesser degree, by all of the countries included in this analysis and possible solutions to these challenges will be discussed in subsequent chapters.

At the same time, intensification of forest management and increased bioenergy feedstock procurement could have adverse effects on ecosystem services, such as soil fertility and biodiversity. Governance that promotes sustainable forest management practices and protects ecosystem services is crucial, in order to ensure the sustainable deployment of forest biomass (see chapter: Environmental Sustainability Aspects of Forest Biomass Mobilisation). In some regions, technical or economic constraints, such as road infrastructure and land access, must also be considered. In addition, certain regions are subject to high rates of natural disturbance, such as fire and insect outbreaks. When it is not possible to reduce or prevent such natural disturbances, the use of salvaged wood for bioenergy production could be a viable option, as the example from Canada shows (Barrette et al., 2015).

REFERENCES

Barrette, J., Thiffault, E., Saint-Pierre, F., Wetzel, S., Duchesne, I., Krigstin, S., 2015. Dynamics of dead tree degradation and shelf-life following natural disturbances: can salvaged trees from boreal forests 'fuel' the forestry and bioenergy sectors? Forestry 88 (3), 275–290.

Bentsen, N.S., Felby, C., 2012. Biomass for energy in the European Union—a review of bioenergy resource assessments. Biotechnol. Biofuels 5 (1), 25.

Berndes, G., Hoogwijk, M., van den Broek, R., 2003. The contribution of biomass in the future global energy supply: a review of 17 studies. Biomass Bioenergy 25 (1), 1–28.

Chum, H., Faaij, A., Moreira, J., Berndes, G., Dhamija, P., Dong, H., Gabrielle, B., Goss Eng, A., Lucht, W., Mapako, M., Masera Cerutti, O., McIntyre, T., Minowa, T., Pingoud, K., 2011. Bioenergy. In: Edenhofer, O., Pichs-Madruga, R., Sokona, Y., Seyboth, K., Matschoss, P., Kadner, S., Zwickel, T., Eickemeier, P., Hansen, G., Schlömer, S., Stechow, C.V. (Eds.), IPCC Special Report on Renewable Energy Sources and Climate Change Mitigation. Cambridge University Press, Cambridge, UK and New York, NY.

Creutzig, F., Ravindranath, N., Berndes, G., Bolwig, S., Bright, R., Cherubini, F., Chum, H., Corbera, E., Delucchi, M., Faaij, A., 2015. Bioenergy and climate change mitigation: an assessment. GCB Bioenergy 7, 916–944.

DEA, 2014. Energy Statistics 2013 (in Danish: Energistatistik 2013). Danish Energy Agency, Danish Ministry of Climate, Energy and Building, Copenhagen, Denmark, Available from: http://www.ens.dk/en/info/facts-figures/energy-statistics-indicators-energy-efficiency/annual-energy-statistics.

DeLucia, E.H., Gomez-Casanovas, N., Greenberg, J.A., Hudiburg, T.W., Kantola, I.B., Long, S.P., Miller, A.D., Ort, D.R., Parton, W.J., 2014. The theoretical limit to plant productivity. Environ. Sci. Technol. 48 (16), 9471–9477.

DOE, 2011. U.S. Billion-Ton Update: Biomass Supply for a Bioenergy and Bioproducts Industry, Oak Ridge National Laboratory. U.S. Department of Energy, Oak Ridge, TN, (ORNL/TM-2011/224). Available from: http://www1.eere.energy.gov/bioenergy/pdfs/billion_ton_update.pdf.

Elbersen, B., Startisky, I., Hengeveld, G., Schelhaas, M.-J., Naeff, H., Böttcher, H., 2012. Atlas of EU biomass potentials: spatially detailed and quantified overview of EU biomass potential taking into account the main criteria determining biomass availability from different sources. Available from: http://www.biomassfutures.eu/work_packages/WP3Supply/D_3_3__Atlas_of_technical_and_economic_biomass_potential_FINAL_Feb_2012.pdf

FAO, 2010. Global Forest Resources Assessment 2010: Main Report. FAO Forestry Paper 163. Food and Agriculture Organization of the United Nations, Rome, Italy, Available from: http://www.fao.org/docrep/013/i1757e/i1757e.pdf.

FAO, 2014. Global Production and Trade of Forest Products in 2013. Food and Agriculture Organization of the United Nations, Rome, Italy, Available from: http://www.fao.org/forestry/statistics/80938/en/.

Field, C.B., Campbell, J.E., Lobell, D.B., 2008. Biomass energy: the scale of the potential resource. Trends Ecol. Evol. 23 (2), 65–72.

Haberl, H., Erb, K.H., Krausmann, F., Gaube, V., Bondeau, A., Plutzar, C., Gingrich, S., Lucht, W., Fischer-Kowalski, M., 2007. Quantifying and mapping the human appropriation of net primary production in earth's terrestrial ecosystems. Proc. Natl. Acad. Sci. 104 (31), 12942–12947.

Hakkila, P., Parikka, M., 2002. Fuel resources from the forest. In: Richardson, J., Björheden, R., Hakkila, P., Lowe, A.T., Smith, C.T. (Eds.), Bioenergy From Sustainable Forestry. Kluwer Academic Publishers, Dordrecht, The Netherlands, pp. 19–48.

Hansen, M.C., Potapov, P.V., Moore, R., Hancher, M., Turubanova, S., Tyukavina, A., Thau, D., Stehman, S., Goetz, S., Loveland, T., 2013. High-resolution global maps of 21st-century forest cover change. Science 342 (6160), 850–853.

Heinimö, J., 2008. Methodological aspects on international biofuels trade: international streams and trade of solid and liquid biofuels in Finland. Biomass Bioenergy 32 (8), 702–716.

Heinsch, F.A., Reeves, M., Votava, P., Kang, S., Milesi, C., Zhao, M., Glassy, J., Jolly, W.M., Loehman, R., Bowker, C.F., 2003. GPP and NPP (MOD17A2/A3) Products NASA MODIS Land Algorithm MOD17 User's Guide. University of Montana, Missoula, MT, pp. 1–57.

Hoogwijk, M., Faaij, A., van den Broek, R., Berndes, G., Gielen, D., Turkenburg, W., 2003. Exploration of the ranges of the global potential of biomass for energy. Biomass Bioenergy 25 (2), 119–133.

IEA, 2014. World Energy Outlook 2014. IEA, Paris, France.

Imhoff, M.L., Bounoua, L., Ricketts, T., Loucks, C., Harriss, R., Lawrence, W.T., 2004. Global patterns in human consumption of net primary production. Nature 429 (6994), 870–873.

IPCC, 2006. 2006 IPCC Guidelines for National Greenhouse Gas Inventories, Prepared by the National Greenhouse Gas Inventories Programme. Institute for Global Environmental Strategies, Kanagawa, Japan.

Ito, A., 2011. A historical meta-analysis of global terrestrial net primary productivity: are estimates converging? Glob. Change Biol. 17 (10), 3161–3175.

Krausmann, F., Erb, K.-H., Gingrich, S., Lauk, C., Haberl, H., 2008. Global patterns of socioeconomic biomass flows in the year 2000: a comprehensive assessment of supply, consumption and constraints. Ecol. Econ. 65 (3), 471–487.

Lauri, P., Havlík, P., Kindermann, G., Forsell, N., Böttcher, H., Obersteiner, M., 2014. Woody biomass energy potential in 2050. Energy Policy 66, 19–31.

Newell, J.P., Simeone, J., 2014. Russia's forests in a global economy: how consumption drives environmental change. Eur. Geogr. Econ. 55 (1), 37–70.

NRCAN, 2014. The State of Canada's Forests—Annual Report 2014. Natural Resources Canada, Canadian Forest Service, Ottawa, Canada, Available from: http://cfs.nrcan.gc.ca/pubwarehouse/pdfs/35713.pdf.

Proskurina, S., Heinimö, J., Mikkilä, M., Vakkilainen, E., 2015. The wood pellet business in Russia with the role of North-West Russian regions: present trends and future challenges. Renew. Sustain. Energy Rev. 51, 730–740.

Rojstaczer, S., Sterling, S.M., Moore, N.J., 2001. Human appropriation of photosynthesis products. Science 294 (5551), 2549–2552.

Running, S.W., 2012. A measurable planetary boundary for the biosphere. Science 337 (6101), 1458–1459.

Slade, R., Bauen, A., Gross, R., 2014. Global bioenergy resources. Nat. Clim. Change 4 (2), 99–105.

Smeets, E.M., Faaij, A.P., Lewandowski, I.M., Turkenburg, W.C., 2007. A bottom-up assessment and review of global bio-energy potentials to 2050. Prog. Energy Combust. Sci. 33 (1), 56–106.

Thiffault, E., Béchard, A., Paré, D., Allen, D., 2014. Recovery rate of harvest residues for bioenergy in boreal and temperate forests: a review. Wiley Interdisc. Rev. Energy Environ. 4 (5), 429–451.

Verbruggen, A., Moomaw, W., Nyboer, J., 2011. I Glossary, acronyms, chemical symbols and prefixes. Intergovernmental Panel Climate Change—Special Report on Renewable Energy Sources Climate Change MitigationCambridge University Press, Cambridge/New York.

Vitousek, P.M., Ehrlich, P.R., Ehrlich, A.H., Matson, P.A., 1986. Human appropriation of the products of photosynthesis. BioScience 36 (6), 368–373.

Wright, D.H., 1990. Human impacts on energy flow through natural ecosystems, and implications for species endangerment. Ambio 19 (4), 189–194.

Chapter 4

Environmental Sustainability Aspects of Forest Biomass Mobilisation

Gustaf Egnell*, David Paré†, Evelyne Thiffault**, Patrick Lamers‡
*Department of Forest Ecology and Management, Swedish University of Agricultural Sciences, Umeå, Sweden; †Canadian Forest Service, Natural Resources Canada, Quebec City, Canada; **Department of Wood and Forest Sciences and Research Centre on Renewable Materials, Laval University, Quebec City, Canada; ‡Idaho National Laboratory, Idaho Falls, ID, United States of America

Highlights

- Bioenergy markets are not likely to become a trigger for large-scale increases in logging and forest area changes in boreal and temperate biomes, under current climate and energy policy regimes. Forest biomass for energy from long-rotation forestry is currently a co-product of forest management for the production of conventional forest products, such as sawlogs, pulp and fibreboard.
- The relevant baseline scenario without a bioenergy market includes wood procurement harvest for conventional forest products—a fact that studies sometimes fail to consider. Enhanced forest biomass procurement translates into an intensification of forest management activities, including changes in silviculture practices, as well as extraction of forest biomass previously left in the forest.
- Forest biomass procurement for energy should be seen as an integral part of silvicultural practices within a forest management strategy. Depending on site conditions, forest biomass procurement can affect other ecosystem services, negatively or positively.
- As such, increased biomass output need not take place at the cost of ecosystem services, if the increased biomass extraction is accompanied by activities that promote regeneration, enhance growth and protect biodiversity, soils and waters.
- The forestry sector needs to start adapting to a future situation where it is expected to provide conventional forest products, biomaterials and bioenergy. To achieve this, good governance mechanisms, such as landscape-level land use planning and science-based improvements of practices, will become increasingly important to ensure sustainable forest product supply chains.

INTRODUCTION

The increasing contribution of forest biomass to the global energy supply is generating concerns about environmental sustainability. These concerns include risks for deterioration of ecosystem services, such as nutrient cycling and soil formation, water purification and flow regulation, biodiversity, as well as carbon sequestration and climate regulation. The environmental consequences of forest biomass procurement need to be well understood, as limits on biomass extraction set by environmental considerations determine how much biomass can be collected (Batidzirai et al., 2013; Chum et al., 2011).

This chapter presents an analysis of the main environmental sustainability aspects of forest biomass procurement from boreal and temperate forests, as well as the driving factors and risks associated with meeting the biomass demand for energy, while providing wood for conventional forest products (ie pulp and paper, sawnwood, composite wood products, panels, etc.). The chapter presents an integrated perspective on the impact of increasing forest biomass for energy use, compared with impacts of forest operations for conventional forest products.

BACKGROUND

Of primary importance in the assessment of the impacts of forest biomass procurement from long-rotation forestry (rotation periods >20 years) is the consideration of a reference state, that is what to compare with, in terms of silvicultural practices and land use, in the absence of a bioenergy market. Woody biomass used for energy is often a by-product of forest operations associated with forest management for the supply of pulpwood and sawn timber. Primary residues, such as branches and tree tops, are available after harvesting operations and secondary residues, such as bark, sawdust and shavings, accumulate during processing operations. Tertiary residues, including end-of-life wood products, such as construction and demolition wood, are all usable for energy (Röser, 2008). Because they differ in characteristics and quality, for example ash content, these fractions are not all equally attractive to conversion facilities and thus do not achieve the same market value. Homogeneous by-product streams, such as sawdust or other mill residues, are the most attractive, for example to the wood pellet production industry. In the absence of such mill residues, large-scale wood pellet production facilities may use whole trees not utilised by conventional forest industries, either due to their characteristics (eg trees of marginal fibre quality, tree size, unwanted species), due to declines in market demand (eg low pulpwood demand leads to declining prices for pulpwood quality logs), or due to increased supply (eg following forest pests, strong winds, or wildfire events).

Emerging bioenergy uses typically benefit from availability of low cost biomass resources, residue streams of various wood processing industries in particular, as well as demolition wood and organic waste. The use of this resource is not likely to compromise sustainability in the forest ecosystem, although there might be some environmental issues linked to transport, storage, conditioning and final use. When secondary and tertiary residues in a region become fully utilised for bioenergy production, branches, tops, stumps, and whole trees are increasingly targeted as a supply. This implies an increase in harvest intensity, both at the site and landscape level, which may have sustainability implications.

ANALYSIS

Land-Use Change

Land-use change (LUC), defined here as a change in land-use category, is driven by several factors, including market economics, land tenure and governance. Potential changes to land uses that can occur as a consequence of increased mobilisation of woody biomass are summarised in Table 4.1. In a situation where forests are managed for sawtimber and pulpwood production, increased demand for forest biomass as energy feedstock rarely results in LUC, but rather incentivises forest land owners to increase revenues from their forests by extracting more biomass, primarily residues as described earlier (eg Abt et al., 2014). Around sources of demand, residue resources may become locally depleted and other local sources of biomass such as pulpwood may be used, unless cheaper biomass can be sourced from longer distances. When sources of demand are close to waterways and railroads, long distance transport is feasible and the abundance of currently unused residues across many regions (eg Dymond et al., 2010, for Canada; Perlack et al., 2011, for the United States; Verkerk et al., 2011b, for the European Union) reduces the likelihood that whole trees and stands be diverted to bioenergy (Table 4.1). Yet, if biomass demand grows beyond what can be supplied based on residue extraction, the relative importance of complementary biomass sources will vary, depending on the competitiveness of the bioenergy sector on agriculture and forestry markets.

In the long term, increased forest growth can support higher biomass output. Many silvicultural treatments are available, including improved site preparation, vegetation control, insect control, genetically improved seedlings, faster growing tree species, fertilisation, irrigation, and so on. Inexpensive silvicultural measures that can be applied on large areas at a fast rate, such as using genetically improved seedling stock, have the highest potential to increase total forest production (Egnell and Björheden, 2013). In addition, new areas can be planted with forests, for example in the South-Eastern United States it is predicted that an enhanced demand for bioenergy would increase timberland area, compared to a situation without such bioenergy demand. However, due to the long-rotation periods, it takes decades before silvicultural measures and forest expansion can make a significant difference in forest production and, thereby, biomass output potential.

Thus, shorter-term responses to biomass demand above possible residue supply require that a larger share of existing forest growth is harvested and/or a larger share of logs are used for bioenergy (eg the cut-off diameter can be increased). This may be considered sustainable, as long as site-specific constraints

TABLE 4.1 Matrix of Potential Changes That May Occur as a Consequence of Greater Woody Biomass Mobilisation for Bioenergy in Situations of Land-Use Change or Land Use Remaining in Original Land Use (Shaded)

From (with no or limited bioenergy market)…	…to (with expanding bioenergy market)		
	Extensive forestry	Intensive forestry	Short-rotation woody crops
Extensive forestry	Higher thinning intensity and increased residue removal: less woody debris at different stand ages; Additional site disturbance; Silviculture measures promoting higher growth: more or less C storage at landscape level. *Likelihood: very common*	Reduced plant diversity; Less dead wood; Greater use of fertiliser and other chemical inputs. *Likelihood: uncommon, because such changes will likely be driven by markets for conventional forest products*	Reduced plant diversity; Less dead wood; Greater use of fertiliser and other chemical inputs. *Likelihood: improbable (physical, technical and financial limitations)*
Intensive forestry	*Unlikely*	Less woody debris at different stand age; Additional site disturbance. *Likelihood: very common*	Less C storage; Habitat complexity reduced; More frequent site disturbance. *Likelihood: uncommon*
Short-rotation woody crops	*Unlikely*	*Likelihood: rare but possible*	No change. *Likelihood: very common*
Agricultural, pasture, or other non-forest land type	Possible environmental benefits. *Likelihood: improbable or very long term*	Increased C storage. *Likelihood: probable (eg SE United States)*	Possible environmental benefits. *Likelihood: possible but extent limited due to other land use demand*

Note: extensive forestry is defined as the practice of forestry on a basis of low operating and investment costs per unit of forest area. Intensive forestry is defined as the practice of forestry to obtain a high level of volume and quality of outturn per unit of area, through the application of the best techniques of silviculture and management. Compared with extensive forestry, intensive forestry requires greater inputs of labour and capital, in terms of quantity, quality, or frequency. Short-rotation woody crops are purpose-grown plantations of fast growing species. The elements included in the matrix need to be interpreted as changes occurring in addition to current forestry activities without a bioenergy market.

are respected and total harvest levels do not exceed forest growth. However, it can slow down the rate of carbon sequestration in the forest landscape and, therefore, be considered non-optimal if forest carbon sinks are relied upon for reaching short-term targets on greenhouse gas (GHG) emissions. Alternatively, biomass output can be ramped up relatively fast by expanding the area under short-rotation woody crops (SRWC) and tree plantations, which may result in direct or indirect forest conversion, depending on where this expansion takes place and what other land uses are displaced.

Agriculture expansion has historically taken place in areas more suitable for agriculture production and remaining forests usually exhibit constraints to the successful implementation of SRWC, such as insufficient soil depth and fertility, stoniness and rough topography. Although some boreal and temperate forest areas under low-intensity management might still support SRWC cultivation, such forest conversion is rare and SRWC have mostly expanded on agriculture land (Table 4.1). Prospects for SRWC cultivation are determined by a range of factors more or less associated with land suitability, including economic performance, market expectations (both bioenergy and food markets), farm structure, risk perception, and aesthetic considerations (McKenney et al., 2011).

Forest management and planning mainly depend on market expectations for conventional forest products, which normally generate greater revenues than bioenergy products. Greater forest biomass use for energy is, therefore, seldom associated with LUC (Abt et al., 2014). However, in some regions, land use is strongly market-driven and competing alternatives like livestock grazing or the expansion of urban areas influence respective LUC decisions. In such cases, both afforestation and deforestation may occur (Table 4.1). However, agriculture expansion is often restricted by unfavourable site conditions in the boreal and large parts of the temperate forest biomes.

Implications of Forest Biomass Removal From Existing Forest Areas

As stated earlier, most boreal and temperate forest biomass that will be used for energy in the coming decades is expected to come from forests managed for sawtimber and pulpwood (not from harvest in undisturbed forests), or from afforestation of non-forested areas. As such, forest biomass procurement to meet bioenergy demand should be analysed as an intensification of forest management, with tree parts and trees being harvested in addition to merchantable wood with market value for conventional forest products, and additional silvicultural practices being put into place to increase the availability of bioenergy feedstocks.

Questions have been raised about the potential environmental impacts of more intensive and frequent harvesting, as well as associated site and landscape disturbance on soil and site productivity, C stocks, water quantity and quality and biodiversity (Lamers et al., 2013). Increased biomass harvesting removes more organic material and nutrients from forests and entails more frequent machinery entries with potential impacts on regeneration, soils and waters. The short-term climate benefits of using biomass from forestry for energy have also been questioned (Ter-Mikaelian et al., 2015), particularly when it comes to coarse woody biomass such as stumps (Pingoud et al., 2012, 2015; Repo et al., 2011; Zanchi et al., 2010) and whole trees (Bernier and Paré, 2013; McKechnie et al., 2010).

The default assumptions that underpin the frames of reference used to assess those questions need to be selected carefully. Vance et al. (2014) offer several examples of default assumptions that often fail to capture the actual forest management operations and current scientific understanding of environmental impacts of forest biomass harvesting, including the following:

- the natural or unmanaged state as the (ideal) reference scenario[1];
- that biomass harvesting removes virtually all residues;
- that biomass, when left on site (ie not recovered for bioenergy), always enhances/maintains soil C, site productivity, water quality, and biodiversity;
- that biomass harvesting is conducted in the absence of operational practices that alleviate site deficiencies, sustain productivity, protect biodiversity, soils and waters; and
- that changes in forest state (ie forest characteristics, such as species composition, tree density, amount of woody debris) are equivalent to changes in forest functions (ie processes performed by a forest ecosystem, such as nutrient and water cycling, providing habitat, etc.).

The default assumption that the environmental characteristics of unmanaged forests are always superior to managed forests is not well supported, and naturalness may be too vague a proxy when it comes to ensuring specific environmental outcomes (Vance et al., 2014). As stated earlier, the proper baseline scenario, when assessing impacts of forest bioenergy systems in the boreal and temperate biomes, is most often forest management for procurement of conventional forest products, not forest conservation.

The data needed for specifying how much biomass can be extracted while sustaining ecosystem functions (soil productivity, biodiversity, surface water quality) are deficient, because the role that biomass plays in ecosystem functioning is highly site-specific (Lamers et al., 2013). Recommendations, for example by Work et al. (2014), suggesting that 45–95 m^3/ha of wood need to be retained to sustain biodiversity, are likely to be specific to the type of stands where the study was conducted (boreal jack pine, in this case). It does not

---

1. The UNEP-SETAC guidelines on methodology for land-use impact assessment requires the definition of a reference situation and mentions three options for the definition of a reference situation: (1) the potential natural vegetation, which 'describes the expected state of mature vegetation in the absence of human intervention'; (2) the (quasi-) natural land cover in each biome/ecoregion, that is 'the natural mix of forests, wetlands, shrubland, grassland, bare area, snow and ice, lakes and rivers', and (3) the current mix of land uses (Koellner et al., 2013).

include quality aspects of the biomass, and it also does not provide the biodiversity-relevant landscape-level assessment. Management of harvested biomass could also be important, since wood-inhabiting insects could lay their eggs in piles of biomass stored in the forest, turning them into insect traps (Victorsson and Jonsell, 2013).

Finally, it has been noted that operational biomass harvests usually report post-harvest changes in leftover biomass levels that are much smaller than the levels applied in most experimental studies. A review of operational recovery rate of harvest residues (tree tops and branches) during clearcutting activities showed that the average recovery rate (solely constrained by operational considerations) in boreal and temperate forests is around 52%, with minimum and maximum values of 4 and 89% (Thiffault et al., 2014). Recovery rates also have a quality and a landscape dimension: high recovery rates at the site level could be counteracted by the quality of retained biomass, as suggested by local conditions and species of concern and the frequency of cuttings where biomass for energy is harvested in the landscape. Nevertheless, a shift in bioenergy policy, a growth in (or a change in access to) bioenergy markets, and upward movements along the technological learning curve (eg improvements in machinery, better training of operators) could increase economic biomass recovery rates and potentially leave unsustainably low biomass levels. This therefore calls for strong science-based governance for biomass procurement (Thiffault et al., 2014).

Soil and Site Productivity

There have been several reviews and meta-analyses looking at the scientific evidence of the effects on soil and site productivity of forest biomass procurement, as an incremental practice to conventional practices of wood procurement for conventional forest products. Overall, it can be concluded that there are no consistent, unequivocal and universal effects of forest biomass harvesting, as effects are site-specific (Vance et al., 2014). Some studies have pointed out that, at the site level, it is harvesting per se (irrelevant to the intensity of biomass removal) that has the greatest ecological impact on some aspects of ecosystem functioning, for example on soil C stocks (Nave et al., 2010) and on invertebrates (Work et al., 2013).

However, negative effects of incremental removal of biomass have also been found, relative to a baseline scenario of harvesting for conventional forest products only (Achat et al., 2015; Thiffault et al., 2011). For example, negative impacts of harvest residue removal (ie tree branches and tops) on soil nutrient pools (eg nitrogen, phosphorus and base cations) and soil acid-base status have been found in the forest floor, but more rarely in the mineral soil. Later in the rotation, impaired nitrogen and/or phosphorus nutrition on whole-tree harvested sites has been shown to reduce tree growth (Thiffault et al., 2011). However, longer-term trials have shown that these reductions in growth rate could be only temporary, as a result of a temporal increase in site productivity when nutrient rich biomass is left on the forest floor (Egnell, 2011). A meta-analysis by Achat et al. (2015) indicates a recovery over time for many response variables, where harvest residues have been removed, including tree growth, soil pH, exchangeable base cations, and nitrogen. Harvest residue removal can also reduce the concentrations of base cations in foliage, but this has not, to date, been shown to affect tree productivity (Thiffault et al., 2011).

One interesting insight from studies on environmental aspects is that biomass procurement practices may mostly influence forest site productivity through influence on the microenvironment of trees, at least during the first years after stand establishment. For example, harvest residue removal can affect seedling growth, either positively or negatively, through its influence on microclimate and competitive vegetation (Trottier-Picard et al., 2014). At the early stage of stand development, the control of the microenvironment (eg through control of competitive vegetation) has been shown to have a consistent, positive influence on forest productivity (Harrington et al., 2013; Holub et al., 2013; Ponder et al., 2012). As such, biomass procurement practices should be viewed as one silvicultural tool among others to manipulate and improve tree growing conditions. This should be taken into account when assessing environmental sustainability of forest biomass procurement. Moreover, application of Best Management Practices (BMP) can help mitigate problems related to soils and site productivity (Neary, 2013).

Surface and Ground Waters

When forest harvest also includes extraction of logging residues for energy markets, this can have either positive or negative impacts on surface and groundwater quality (Laudon et al., 2011). Positive impacts include eutrophication reduction, where removal of nitrogen-rich logging residues can reduce nitrate leaching to groundwater surrounding surface waters (Slesak et al., 2009). This could be relevant in areas with high nitrogen deposition and leaching rates (Staaf and Olsson, 1994). The increase in soil pH and base saturation, following harvest, tends to be slightly lower if nutrient rich logging residues are also harvested (Staaf and Olsson, 1991), but there seems to be a recovery over time (Achat et al., 2015). Thus, in the short-term, lower pH and base saturation may change the quality of the water reaching surface waters (Stevens et al., 1995).

In terms of negative impacts, however, the largest threat for waters comes as a result of sediment from rill and gully erosion and slope failures that direct runoff water into surface waters. Fine textured mineral soil, organic material and toxic elements like methyl mercury, transported by surface runoff, can reduce water quality (Eklöf et al., 2012). These problems are common also when stemwood for conventional forest products is harvested. But the risk of soil damage increases as a result of more machine traffic, steeper slopes, fragile soils and the fact that harvested logging residues cannot be used to reinforce the strip roads (Eliasson and Wästerlund, 2007). Furthermore, stump harvest for the energy market, as currently practiced in Scandinavia, causes damage where the stumps are extracted (Kataja-aho et al., 2012). The potential to counteract these effects is, however, high with BMP such as wood-ash recycling to counteract acidification (Piirainen et al., 2013), buffer zones adjacent to surface waters (Neary, 2013), and hydrological mapping and planning (Kuglerová et al., 2014). This need for hydrological planning could be further strengthened by a changing climate with shorter periods of ground frost and, in some areas, predicted increases in precipitation.

Collection of forest harvesting residues for bioenergy feedstocks generally does not cause adverse effects on water quantity, unless erosion and sediment loss inhibit surface runoff (Neary and Koestner, 2012). Deterioration of water quality can reduce the amount of quality water suitable for municipal water supplies (Neary et al., 2011).

Biodiversity

Looking at site-level impacts of forest biomass procurement on biodiversity, a meta-analysis on the topic has shown that diversity and abundance of birds and invertebrates can be substantially and consistently lower in treatments with lower amounts of downed woody debris and/or standing snags. Effects for other taxa were not as large, were based on fewer studies, and varied among studies. Another meta-analysis looking at the impacts of forest thinning treatments, which is also a forest biomass procurement option, found that they had generally positive or neutral effects on diversity and abundance across all taxa, although thinning intensity and the type of thinning conducted may drive, at least partially, the magnitude of response (Verschuyl et al., 2011).

The effects of the intensification of forest management should also be considered at the landscape-level, especially for impacts on biodiversity. However, empirical evidence is much harder to gather when studying landscape effects, as opposed to site-level effects, since true replication in space and/or time is often impossible to achieve. Therefore, landscape-level studies on forest procurement practices often rely on modelling, with the caveats that such work carries. For example, a modelling study by Verkerk et al. (2011a) suggested that intensified forest biomass removal would affect significantly the amount of deadwood (including both downed debris and standing dead trees) at country level, for some European countries, with potential impacts on biodiversity. On the other hand, a modelling study on biodiversity of mushrooms, lichens and beetles concluded that, although populations will go down as a result of substrate reduction, extraction of 70% of Norway spruce branches and tops on 50% of the clearcuts in Sweden constitutes a minor contribution to the regional extinction risks, since red-listed species in Sweden are not associated with spruce logging residues (Dahlberg et al., 2011). Also, another modelling study in Europe suggested that protected forests—a segregative approach in which ecosystems and species are conserved when they are unlikely to survive in intensively managed areas—do not affect wood supply, given the current demand for wood in Europe. However, if demand for wood from European forests for material and energy use increases significantly, then there might be pressure to 'unlock' biomass volumes located in protected areas (Verkerk et al., 2014). Therefore, for preservation purposes, it is important to pay special attention to dead and living trees and stands with high biodiversity potential that, without a bioenergy market, would have been left uncut. It is also important to keep in mind that markets for biomass can be an integrated part in the management efforts to strengthen biodiversity in both production forests and in protected forests, for example by removing exotic and invasive plants and trees, or keeping biodiverse woodlands/grasslands open (Van Meerbeek et al., 2015).

Concerns of landscape-level impacts of intensification of forest management and biomass procurement on biodiversity also echo concerns about the expansion of forest plantations, at the expense of primary or natural forests. Plantation forestry has acquired a bad reputation, mainly resulting from intensive, large-scale monocultures. However, many other options exist that could also be used to meet biomass demand, along with other ecosystem objectives. Plantation forests need not be 'biological deserts'; although they differ in their absolute biodiversity value, relative to natural forests (Barlow et al., 2007), human-modified forests, such as plantations, can still have a conservation value when they support threatened or evolutionarily isolated biodiversity (Daily, 2001; Lindenmayer and Franklin, 2002). Well-designed, multi-purposed plantations can help mitigate climate change through direct C absorption, while simultaneously protecting remaining natural forests through increased productivity at landscape level (Messier et al., 2013). The use of protected areas in conjunction with sustainable forest management (including establishment of plantations) in the landscape is one key to ensure that forest landscapes can provide forest products, while biodiversity is still conserved and C storage capacity is not degraded (Elbakidze et al., 2013).

Carbon and Climate Change

Bioenergy is part of the terrestrial C cycle: the CO_2 emitted due to bioenergy use was previously sequestered from the atmosphere and will be sequestered again, if the bioenergy system is managed sustainably. In contrast, the use of fossil fuels transfers CO_2 from a stable geologic storage into the atmosphere (IPCC, 2014). Life-cycle analysis and C footprint studies of bioenergy systems have commonly excluded the C sequestration and C emissions associated with plant growth and subsequent bioenergy use. This so-called 'C neutrality assumption' of bioenergy is debated; in the case of forest-based bioenergy, it is the timing of net GHG savings that is in focus. For example, if logging residues are collected and used for bioenergy, their C content will be emitted to the atmosphere much more rapidly than under a reference scenario where these residues are left in the forest to decay. This difference in timing of C emissions between the reference and bioenergy scenarios is a critical factor determining whether forest bioenergy is found to contribute positively to net GHG savings or not, at least in the near term.

The C accounting of bioenergy in national and international GHG emission schemes follows that applied in the Kyoto Protocol. Here, emissions from bioenergy systems are excluded in the energy sector, on the assumption that any loss in forest C stock due to biomass harvest is treated as an emission in the land-use sector. The reason for this is to avoid double-counting (IPCC, 2006). However, in the context of energy sector emissions, this accounting gives bioenergy the perception of 'carbon neutrality'; an assumption that has been widely criticised (Johnson, 2009; Searchinger et al., 2009). The two main points of critique include the fact that the combustion of biomass emits the C content to the atmosphere immediately, whereas, if left in the forest, it would decompose over time, with the C partly accumulating in the soil and partly released into the atmosphere. Secondly, the energy output per unit of C emitted is lower for biomass than for fossil alternatives. This creates a payback time that describes the time lag between the point of biomass harvest and C release as the biomass is used for energy, and the point in time when the same (absolute) or a reference (relative) forest and energy system reach the same C concentration levels. This implies that the initial C loss caused by biomass procurement is paid back, for example via C sequestration in new vegetation and avoided C emissions by displacing fossil fuels. The payback time essentially shows the delay until C neutrality is reached. This time difference has caused debate as to whether bioenergy is able to help achieve near-term GHG reduction targets (Cowie et al., 2013).

The fate of woody debris, harvest residues, snags and trees in the baseline scenario without biomass procurement (ie sources of biomass that could otherwise be removed for bioenergy) has particular relevance for calculations of CO_2 emission mitigation benefits of bioenergy from forest biomass. The decomposition rate of leftover biomass and the fate of its C content (and whether and to what extent it contributes to the soil C pool and site productivity) are essential information that help to define the reference scenario against which the effects of biomass procurement for bioenergy production are assessed. Estimates of residence times for leftover biomass vary from 2 years up to 500 years, depending on biomass characteristics (eg species and diameter) and position (eg close contact to the ground, suspended in the air, etc.), climate and other site conditions and management practices (Vance et al., 2014). However, most field studies suggest that, in a baseline scenario without biomass procurement, leftover debris does not add significant amounts of C to soil pools in the long run, relative to scenarios where debris is removed, except in some cases (eg boreal sites on coarse-textured soils) (Nave et al., 2010; Thiffault et al., 2011).

Nevertheless, the payback time is assumed to be particularly long for coarse woody biomass like stumps and stemwood from long-rotation forestry, since they potentially continue to grow and/or decompose more slowly than tree tops and branches, if left in the forest (Repo et al., 2012). It has been argued, therefore, that stemwood should not be used as a source for bioenergy (Agostini et al., 2013). In the larger context of climate change mitigation, however, this conclusion appears premature. Mitigating global warming is a long-term, rather than a short-term objective. Therefore, short term increases in GHG emissions, as a result of using biomass from long-rotation forestry as an energy source, will eventually turn into long-term GHG reductions, as compared to fossil alternatives (Dehue, 2013). Transitioning to an energy system devoid of fossil fuels and based on renewable sources might require some short-term increases in emissions, as long as they are part of a larger transition plan, which should include preparing the forestry sector for a future situation where it is expected to provide biomaterials and bioenergy. Since climate change mitigation is a long-term target, it is important that policymakers are aware of the long-term benefits and do not base their policies solely on short-term GHG emission targets.

For example, in forest stands in which there is a high proportion of trees with limited value for conventional forest products (due to tree species, size or fibre quality), the proportion of high-value sawlogs that could be harvested is small, which affects the financial viability of the forest value chain and could paralyse forest management activities. Adding bioenergy to the basket of products that can be sourced from a given stand may increase the profitability of the overall forest operations and create incentives for forest management by providing an outlet for lower-quality fibre. This will provide benefits to the whole forest sector and ensures the greatest benefits (in terms of GHG savings) by creating a flow of forest-based products with often high substitution and GHG mitigation benefits (Sathre and O'Connor, 2010; Sikkema et al., 2014). An increase in the use of wood will result in an increased residual stream that could be used for bioenergy. Furthermore, it may increase foresters' belief in future markets giving them incentives to invest in measures to increase forest productivity (Bellassen and Luyssaert, 2014). Analyses including the full suite of forest products do indeed show the benefits of using wood from sustainable forestry from a climate warming mitigation perspective (Lundmark et al., 2014; Smyth et al., 2014).

The climate benefits of using wood to substitute materials resulting in large GHG emissions has initiated a debate on regulating the use of stemwood in Europe, according to the cascading principle, where combustion of biomass from stemwood would only be eligible after it has been used in other forest products. However, such a constraint would prevent stemwood that have no industrial uses (such as wood of inferior quality left out in harvest operations, or residual stemwood generated by natural disturbance events) being used in bioenergy. As such, the cascading principle may not be a viable strategy to optimise environmental benefits from forestry including GHG savings. On the contrary, a bioenergy market could help the forest sector to adapt to natural disturbances, since it has the potential to take care of large quantities of low quality stemwood (Barrette et al., 2015; Mansuy et al., 2015). Thus, excluding certain types of bioenergy feedstocks, such as stemwood, solely based on their theoretical C payback time might not be the best option from a forestry and climate perspective. However, policies to increase markets for construction wood to replace steel and concrete could be a way forward; an increased use of wood will result in an increased residue stream for the energy market and most likely to more investments in forest growth that will increase sustainable annual harvest levels from a feedstock supply perspective (see chapter: Quantifying Forest Biomass Mobilisation Potential in the Boreal and Temperate Biomes).

SUMMARY AND CONCLUSIONS

Some studies foresee vast increases in harvest operations, as a response to a bioenergy market demand, based on assumptions about how much of the world's energy use should be covered by biomass from forests (Schulze et al., 2012). These studies are correct in their conclusions that if forests were to produce a very large part of global primary energy use, this would have major impacts on ecosystems globally. But using empirical observations of countries from the boreal and temperate biomes with different levels of forest biomass mobilisation, it seems that these studies are incorrect in their assumptions that bioenergy markets are a trigger for large-scale increases in logging and land-use changes. Forest biomass for energy from long-rotation forestry is—and will most likely continue to be—a co-product of forest management for conventional forest products. The main feedstocks will primarily consist of residue streams in the forest and in the forest industry, providing benefits to the whole forest sector.

Nevertheless, the principles of protection and sustainability should remain the same whether forests are managed for conventional forest products with or without biomass for bioenergy. Adaptation of forest management strategies is needed where evidence suggests that the incremental removal of biomass or other forms of intensive management may not be sustainable. Silvicultural practices, such as fertilisation, competition control and soil preparation need to be available as options to manage the microenvironment and tree growing conditions and to prevent or mitigate negative impacts. Importantly, biomass removal should also be seen as an integral part of these silvicultural practices. Moreover, landscape management practices/regulations should be enforced to ensure that sufficient biodiversity-important features, such as dead wood, aging stands, corridors, etc. are preserved. Greater emphasis should also be placed on linking practices to functional values of interest, whether or not they deviate from some natural state or standard. As described by the concept of adaptive forest management, this requires a proper mechanism of ecological monitoring of harvesting operations, scientific field testing and modelling that interact to produce better knowledge that could then help improve practices (Thiffault et al., 2010).

Chum et al. (2011) developed four different storylines for biomass deployment on a global scale, by 2050. Adapting these storylines to the specific context of forest biomass (Fig. 4.1) suggests that high biomass deployment levels can be reached in several different ways. Good governance mechanisms, such as landscape-level land use planning and science-based improvements of practices (including increasing the awareness of the forestry sector for a future situation where it is expected to provide conventional forest products, biomaterials and bioenergy) is needed to ensure mobilisation of sustainable forest biomass supply chains. This also suggests that high levels of sustainable mobilisation would require high levels of globalisation of supply chains, for example through:

- more fluid international trade routes (see chapter: Challenges and Opportunities for International Trade in Forest Biomass);

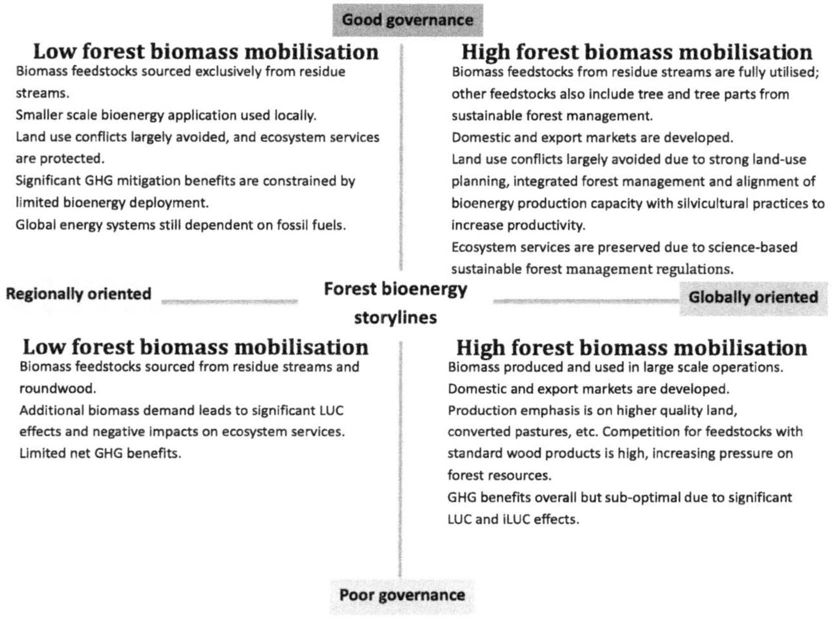

Good governance

Low forest biomass mobilisation

Biomass feedstocks sourced exclusively from residue streams.

Smaller scale bioenergy application used locally.

Land use conflicts largely avoided, and ecosystem services are protected.

Significant GHG mitigation benefits are constrained by limited bioenergy deployment.

Global energy systems still dependent on fossil fuels.

High forest biomass mobilisation

Biomass feedstocks from residue streams are fully utilised; other feedstocks also include tree and tree parts from sustainable forest management.

Domestic and export markets are developed.

Land use conflicts largely avoided due to strong land-use planning, integrated forest management and alignment of bioenergy production capacity with silvicultural practices to increase productivity.

Ecosystem services are preserved due to science-based sustainable forest management regulations.

Regionally oriented ———————— **Forest bioenergy** ———————— **Globally oriented**

storylines

Low forest biomass mobilisation

Biomass feedstocks sourced from residue streams and roundwood.

Additional biomass demand leads to significant LUC effects and negative impacts on ecosystem services.

Limited net GHG benefits.

High forest biomass mobilisation

Biomass produced and used in large scale operations.

Domestic and export markets are developed.

Production emphasis is on higher quality land, converted pastures, etc. Competition for feedstocks with standard wood products is high, increasing pressure on forest resources.

GHG benefits overall but sub-optimal due to significant LUC and iLUC effects.

Poor governance

FIGURE 4.1   Forest bioenergy storylines, and associated levels of forest biomass mobilisation. *(Source: Adapted from Chum et al. (2011).)*

- globally harmonised sustainability requirements and certification systems (see chapter: Challenges and Opportunities for International Trade in Forest Biomass);
- more technology and knowledge transfer across regions, that is bioenergy late-comers could leap-frog towards state-of-the-art technology and practices developed in countries with a longer history of bioenergy deployment (see chapter: Constraints and Success Factors for Woody Biomass Energy Systems in Two Countries With Minimal Bioenergy Sectors).

However, this does not imply that this scenario is appropriate, acceptable, desirable or even possible in all conditions in every region of the globe; local stakeholders need to have a say on the development of their own environment.

REFERENCES

Abt, K.L., Abt, R.C., Galik, C.S., Skog, K.E., 2014. Effect of Policies on Pellet Production and Forests in the US South: A Technical Document Supporting the Forest Service Update of the 2010 RPA Assessment. USDA, Washington, DC, Available from: http://www.srs.fs.usda.gov/pubs/gtr/gtr_srs202.pdf

Achat, D., Deleuze, C., Landmann, G., Pousse, N., Ranger, J., Augusto, L., 2015. Quantifying consequences of removing harvesting residues on forest soils and tree growth—a meta-analysis. Forest Ecol. Manag. 348, 124–141.

Agostini, A., Giuntoli, J. and Boulamanti, A., 2013. Carbon Accounting of Forest Bioenergy. Available from: http://iet.jrc.ec.europa.eu/bf-ca/sites/bf-ca/files/files/documents/eur25354en_online-final.pdf

Barlow, J., Gardner, T.A., Araujo, I.S., Ávila-Pires, T.C., Bonaldo, A.B., Costa, J.E., Esposito, M.C., Ferreira, L.V., Hawes, J., Hernandez, M.I., 2007. Quantifying the biodiversity value of tropical primary, secondary, and plantation forests. Proc. Natl. Acad. Sci. 104 (47), 18555–18560.

Barrette, J., Thiffault, E., Saint-Pierre, F., Wetzel, S., Duchesne, I., Krigstin, S., 2015. Dynamics of dead tree degradation and shelf-life following natural disturbances: can salvaged trees from boreal forests 'fuel' the forestry and bioenergy sectors? Forestry 88 (3), 275–290.

Batidzirai, B., Mignot, A., Schakel, W., Junginger, H., Faaij, A., 2013. Biomass torrefaction technology: techno-economic status and future prospects. Energy 62, 196–214.

Bellassen, V., Luyssaert, S., 2014. Carbon sequestration: managing forests in uncertain times. Nature 506 (7487), 153–155.

Bernier, P., Paré, D., 2013. Using ecosystem CO_2 measurements to estimate the timing and magnitude of greenhouse gas mitigation potential of forest bioenergy. GCB Bioenergy 5 (1), 67–72.

Chum, H., Faaij, A., Moreira, J., Berndes, G., Dhamija, P., Dong, H., Gabrielle, B., Goss Eng, A., Lucht, W., Mapako, M., Masera Cerutti, O., McIntyre, T., Minowa, T., Pingoud, K., 2011. Bioenergy. In: Edenhofer, O., Pichs-Madruga, R., Sokona, Y., Seyboth, K., Matschoss, P., Kadner, S., Zwickel, T., Eickemeier, P., Hansen, G., Schlömer, S., Stechow, C.V. (Eds.), IPCC Special Report on Renewable Energy Sources and Climate Change Mitigation. Cambridge University Press, Cambridge, UK and New York, NY.

Cowie, A., Berndes, G., Smith, T., 2013. On the timing of greenhouse gas mitigation benefits of forest-based bioenergy. IEA Bioenergy Executive Committee Statement, 2013, 4.

Dahlberg, A., Thor, G., Allmér, J., Jonsell, M., Jonsson, M., Ranius, T., 2011. Modelled impact of Norway spruce logging residue extraction on biodiversity in Sweden. Can. J. Forest Res. 41 (6), 1220–1232.

Daily, G.C., 2001. Ecological forecasts. Nature 411 (6835), 245.

Dehue, B., 2013. Implications of a 'carbon debt' on bioenergy's potential to mitigate climate change. Biofuels Bioprod. Biorefin. 7 (3), 228–234.

Dymond, C.C., Titus, B.D., Stinson, G., Kurz, W.A., 2010. Future quantities and spatial distribution of harvesting residue and dead wood from natural disturbances in Canada. Forest Ecol. Manag. 260 (2), 181–192.

Egnell, G., 2011. Is the productivity decline in Norway spruce following whole-tree harvesting in the final felling in boreal Sweden permanent or temporary? Forest Ecol. Manag. 261 (1), 148–153.

Egnell, G., Björheden, R., 2013. Options for increasing biomass output from long-rotation forestry. Wiley Interdisc. Rev. Energy Environ. 2 (4), 465–472.

Eklöf, K., Kraus, A., Weyhenmeyer, G., Meili, M., Bishop, K., 2012. Forestry influence by stump harvest and site preparation on methylmercury, total mercury and other stream water chemistry parameters across a boreal landscape. Ecosystems 15 (8), 1308–1320.

Elbakidze, M., Angelstam, P., Sobolev, N., Degerman, E., Andersson, K., Axelsson, R., Höjer, O., Wennberg, S., 2013. Protected area as an indicator of ecological sustainability? A century of development in Europe's boreal forest. Ambio 42 (2), 201–214.

Eliasson, L., Wästerlund, I., 2007. Effects of slash reinforcement of strip roads on rutting and soil compaction on a moist fine-grained soil. Forest Ecol. Manag. 252 (1–3), 118–123.

Harrington, T.B., Slesak, R.A., Schoenholtz, S.H., 2013. Variation in logging debris cover influences competitor abundance, resource availability, and early growth of planted Douglas-fir. Forest Ecol. Manag. 296, 41–52.

Holub, S.M., Terry, T.A., Harrington, C.A., Harrison, R.B., Meade, R., 2013. Tree growth ten years after residual biomass removal, soil compaction, tillage, and competing vegetation control in a highly-productive Douglas-fir plantation. Forest Ecol. Manag. 305, 60–66.

IPCC, 2006. 2006 IPCC Guidelines for National Greenhouse Gas Inventories, Prepared by the National Greenhouse Gas Inventories Programme. Institute for Global Environmental Strategies, Kanagawa, Japan.

IPCC, 2014. Climate Change 2014: Mitigation of Climate Change. Contribution of Working Group III to the Fifth Assessment Report of the Intergovernmental Panel on Climate Change. Cambridge University Press, Cambridge, UK and New York, NY.

Johnson, E., 2009. Goodbye to carbon neutral: getting biomass footprints right. Environ. Impact Assess. Rev. 29 (3), 165–168.

Kataja-aho, S., Smolander, A., Fritze, H., Norrgård, S., Haimi, J., 2012. Responses of soil carbon and nitrogen transformations to stump removal. Silva Fennica 46 (2), 169–179.

Koellner, T., de Baan, L., Beck, T., Brandão, M., Civit, B., Margni, M., Milà i Canals, L., Saad, R., de Souza, D., Müller-Wenk, R., 2013. UNEP-SETAC guideline on global land use impact assessment on biodiversity and ecosystem services in LCA. Int. J. Life Cycle Assess. 18 (6), 1188–1202.

Kuglerová, L., Ågren, A., Jansson, R., Laudon, H., 2014. Towards optimizing riparian buffer zones: Ecological and biogeochemical implications for forest management. Forest Ecol. Manag. 334, 74–84.

Lamers, P., Thiffault, E., Paré, D., Junginger, M., 2013. Feedstock specific environmental risk levels related to biomass extraction for energy from boreal and temperate forests. Biomass Bioenergy 55 (8), 212–226.

Laudon, H., Sponseller, R.A., Lucas, R.W., Futter, M.N., Egnell, G., Bishop, K., Ågren, A., Ring, E., Högberg, P., 2011. Consequences of more intensive forestry for the sustainable management of forest soils and waters. Forests 2 (1), 243–260.

Lindenmayer, D.B., Franklin, J.F., 2002. Conserving Forest Biodiversity: A Comprehensive Multi-scaled Approach. Island Press, Washington, DC.

Lundmark, T., Bergh, J., Hofer, P., Lundström, A., Nordin, A., Poudel, B.C., Sathre, R., Taverna, R., Werner, F., 2014. Potential roles of Swedish forestry in the context of climate change mitigation. Forests 5 (4), 557–578.

Mansuy, N., Thiffault, E., Lemieux, S., Manka, F., Paré, D., Lebel, L., 2015. Sustainable biomass supply chains from salvage logging of fire-killed stands: a case study for wood pellet production in eastern Canada. Appl. Energy 154, 62–73.

McKechnie, J., Colombo, S., Chen, J., Mabee, W., MacLean, H.L., 2010. Forest bioenergy or forest carbon? Assessing trade-offs in greenhouse gas mitigation with wood-based fuels. Environ. Sci. Technol. 45 (2), 789–795.

McKenney, D.W., Yemshanov, D., Fraleigh, S., Allen, D., Preto, F., 2011. An economic assessment of the use of short-rotation coppice woody biomass to heat greenhouses in southern Canada. Biomass Bioenergy 35 (1), 374–384.

Messier, C., Puettmann, K.J., Coates, K.D., 2013. Managing Forests as Complex Adaptive Systems: Building Resilience to the Challenge of Global Change. Routledge, London.

Nave, L.E., Vance, E.D., Swanston, C.W., Curtis, P.S., 2010. Harvest impacts on soil carbon storage in temperate forests. Forest Ecol. Manag. 259 (5), 857–866.

Neary, D.G., 2013. Best management practices for forest bioenergy programs. Wiley Interdisc. Rev. Energy Environ. 2 (6), 614–632.

Neary, D.G., Koestner, K.A., 2012. Forest bioenergy feedstock harvesting effects on water supply. Wiley Interdisc. Rev. Energy Environ. 1 (3), 270–284.

Neary, D.G., Smethurst, P.J., Baillie, B., Petrone, K.C., 2011. Water quality, biodiversity and codes of practice in relation to harvesting forest plantations in streamside management zones. CSIRO, National Research Flagships. 101 p. Available at: http://www.fs.fed.us/rm/pubs_other/rmrs_2011_neary_d004.pdf

Perlack, R.D., Eaton, L.M., Turhollow Jr, A.F., Langholtz, M.H., Brandt, C.C., Downing, M.E., Graham, R.L., Wright, L.L., Kavkewitz, J.M. and Shamey, A.M., 2011. US Billion-Ton Update: Biomass Supply for a Bioenergy and Bioproducts Industry. Available from: http://www1.eere.energy.gov/bioenergy/pdfs/billion_ton_update.pdf

Piirainen, S., Domisch, T., Moilanen, M., Nieminen, M., 2013. Long-term effects of ash fertilization on runoff water quality from drained peatland forests. Forest Ecol. Manag. 287, 53–66.

Pingoud, K., Ekholm, T., Savolainen, I., 2012. Global warming potential factors and warming payback time as climate indicators of forest biomass use. Mitig. Adapt. Strateg. Glob. Change 17 (4), 369–386.

Pingoud, K., Ekholm, T., Soimakallio, S., Helin, T., 2016. Carbon balance indicator for forest bioenergy scenarios. GCB Bioenergy 8: 171–182. doi:10.1111/gcbb.12253.

Ponder, F., Fleming, R.L., Berch, S., Busse, M.D., Elioff, J.D., Hazlett, P.W., Kabzems, R.D., Kranabetter, J.M., Morris, D.M., Page-Dumroese, D., 2012. Effects of organic matter removal, soil compaction and vegetation control on 10th year biomass and foliar nutrition: LTSP continent-wide comparisons. Forest Ecol. Manag. 278, 35–54.

Repo, A., Tuomi, M., Liski, J., 2011. Indirect carbon dioxide emissions from producing bioenergy from forest harvest residues. GCB Bioenergy 3 (2), 107–115.

Repo, A., Känkänen, R., Tuovinen, J.-P., Antikainen, R., Tuomi, M., Vanhala, P., Liski, J., 2012. Forest bioenergy climate impact can be improved by allocating forest residue removal. GCB Bioenergy 4 (2), 202–212.

Röser, D., Asikainen, A., Raulund-Rasmussen, K., Stupak, I. (Eds.), 2008. Sustainable use of forest biomass for energy: a synthesis with focus on the baltic and nordic region, vol. 12, Springer Science & Business Media, p. 261.

Sathre, R., O'Connor, J., 2010. Meta-analysis of greenhouse gas displacement factors of wood product substitution. Environ. Sci. Policy 13 (2), 104–114.

Schulze, E.D., Körner, C., Law, B.E., Haberl, H., Luyssaert, S., 2012. Large-scale bioenergy from additional harvest of forest biomass is neither sustainable nor greenhouse gas neutral. GCB Bioenergy 4 (6), 611–616.

Searchinger, T.D., Hamburg, S.P., Melillo, J., Chameides, W., Havlik, P., Kammen, D.M., Likens, G.E., Lubowski, R.N., Obersteiner, M., Oppenheimer, M., 2009. Fixing a critical climate accounting error. Science 326 (5952), 527.

Sikkema, R., Junginger, M., van Dam, J., Stegeman, G., Durrant, D., Faaij, A., 2014. Legal harvesting, sustainable sourcing and cascaded use of wood for bioenergy: their coverage through existing certification frameworks for sustainable forest management. Forests 5 (9), 2163–2211.

Slesak, R.A., Schoenholtz, S.H., Harrington, T.B., Strahm, B.D., 2009. Dissolved carbon and nitrogen leaching following variable logging-debris retention and competing-vegetation control in Douglas-fir plantations of western Oregon and Washington. Can. J. Forest Res. 39 (8), 1484–1497.

Smyth, C., Stinson, G., Neilson, E., Lemprière, T., Hafer, M., Rampley, G., Kurz, W., 2014. Quantifying the biophysical climate change mitigation potential of Canada's forest sector. Biogeosciences 11 (13), 3515–3529.

Staaf, H., Olsson, B.A., 1991. Acidity in four coniferous forest soils after different harvesting regimes of logging slash. Scand. J. Forest Res. 6 (1–4), 19–29.

Staaf, H., Olsson, B.A., 1994. Effects of slash removal and stump harvesting on soil water chemistry in a clearcutting in SW Sweden. Scand. J. Forest Res. 9 (1–4), 305–310.

Stevens, P.A., Norris, D.A., Williams, T.G., Hugues, S., Durrant, D.W.H., Anderson, M.A., Weatherley, N.S., Hornung, M., Woods, C., 1995. Nutrient losses after clearfelling in Beddgelert Forest: a comparison of the effects of conventional and whole-tree harvest on soil water chemistry. Forestry 68 (2), 115–131.

Ter-Mikaelian, M.T., Colombo, S.J., Chen, J., 2015. The burning question: does forest bioenergy reduce carbon emissions? A review of common misconceptions about forest carbon accounting. J. Forestry 113 (1), 57–68.

Thiffault, E., Béchard, A., Paré, D., Allen, D., 2014. Recovery rate of harvest residues for bioenergy in boreal and temperate forests: a review. Wiley Interdisc. Rev. Energy Environ. 4 (5), 429–451.

Thiffault, E., Hannam, K.D., Paré, D., Titus, B.D., Hazlett, P.W., Maynard, D.G., Brais, S., 2011. Effects of forest biomass harvesting on soil productivity in boreal and temperate forests—a review. Environ. Rev. 19, 278–309.

Thiffault, E., Paré, D., Brais, S., Titus, B.D., 2010. Intensive biomass removals and site productivity in Canada: a review of relevant issues. Forestry Chron. 86 (1), 36–42.

Trottier-Picard, A., Thiffault, E., DesRochers, A., Paré, D., Thiffault, N., Messier, C., 2014. Amounts of logging residues affect planting microsites: a manipulative study across northern forest ecosystems. Forest Ecol. Manag. 312, 203–215.

Van Meerbeek, K., Appels, L., Dewil, R., Calmeyn, A., Lemmens, P., Muys, B., Hermy, M., 2015. Biomass of invasive plant species as a potential feedstock for bioenergy production. Biofuels Bioprod. Biorefin. 9 (3), 273–282.

Vance, E.D., Aust, W.M., Strahm, B.D., Froese, R.E., Harrison, R.B., Morris, L.A., 2014. Biomass harvesting and soil productivity: is the science meeting our policy needs? Soil Sci. Soc. Am. J. 78 (S1), S95–S104.

Verkerk, P., Lindner, M., Zanchi, G., Zudin, S., 2011a. Assessing impacts of intensified biomass removal on deadwood in European forests. Ecol. Indicators 11 (1), 27–35.

Verkerk, P.J., Anttila, P., Eggers, J., Lindner, M., Asikainen, A., 2011b. The realisable potential supply of woody biomass from forests in the European Union. Forest Ecol. Manag. 261 (11), 2007–2015.

Verkerk, P., Mavsar, R., Giergiczny, M., Lindner, M., Edwards, D., Schelhaas, M., 2014. Assessing impacts of intensified biomass production and biodiversity protection on ecosystem services provided by European forests. Ecosyst. Serv. 9, 155–165.

Verschuyl, J., Riffell, S., Miller, D., Wigley, T.B., 2011. Biodiversity response to intensive biomass production from forest thinning in North American forests—a meta-analysis. Forest Ecol. Manag. 261 (2), 221–232.

Victorsson, J., Jonsell, M., 2013. Ecological traps and habitat loss, stump extraction and its effects on saproxylic beetles. Forest Ecol. Manag. 290, 22–29.

Work, T.T., Brais, S., Harvey, B.D., 2014. Reductions in downed deadwood from biomass harvesting alter composition of spiders and ground beetle assemblages in jack-pine forests of Western Quebec. Forest Ecol. Manag. 321, 19–28.

Work, T.T., Klimaszewski, J., Thiffault, E., Bourdon, C., Pare, D., Bousquet, Y., Venier, L., Titus, B., 2013. Initial responses of rove and ground beetles (Coleoptera, Staphylinidae, Carabidae) to removal of logging residues following clearcut harvesting in the boreal forest of Quebec, Canada. ZooKeys 258, 31.

Zanchi, G., Pena, N., Bird, N., 2010. The Upfront Carbon Debt of Bioenergy. Joanneum, Graz, Austria, Available from: http://www.birdlife.org/eu/pdfs/Bioenergy_Joanneum_Research.pdf

Chapter 5

Challenges and Opportunities of Logistics and Economics of Forest Biomass

Antti Asikainen, Tanja Ikonen, Johanna Routa
Natural Resources Institute Finland (LUKE), Joensuu, Finland

Highlights

- The integration of the conventional wood products industry and the energy industry is the corner stone of a market-driven replacement of fossil fuels.
- Although the productivity of machinery is important, feedstock quality management is particularly critical for ensuring the success of forest biomass supply chains.
- Energy yields per unit of delivered biomass can be maximised through careful establishment and location of storage, prediction and measurement of changing moisture content, as well as the ability to match supply with demand.
- Biomass projects should target areas where market-driven competitiveness is best and where economic sustainability can be achieved with modest incentives.

INTRODUCTION

The forest biomass supply chain generally consists of three inter-related phases: feedstock production, feedstock-to-bioenergy conversion and energy distribution to the customer. The first phase requires: identifying and purchasing potential biomass resources; harvesting, forwarding and comminuting the biomass; transporting it to the plant; and receiving and handling it at the plant site. This chapter describes the logistical, economic and fuel quality-related challenges that face forest biomass supply chains when competing with other feedstocks or energy markets. In this chapter, the technology, economics and overall competitiveness of forest biomass as an energy feedstock are reviewed, with a special emphasis on the boreal and temperate biomes and on the domestic supply to local and regional end-users.

Mobilisation of Forest Bioenergy in the Boreal and Temperate Biomes

MATCHING SUPPLY TO DEMAND

Spatial and Temporal Challenges of Feedstock Supply

The global annual use of wood amounts to 3.6×10^9 m³: energy production consumes 1.9×10^9 m³ and industrial processing (eg timber, pulp and paper) consumes the remaining 1.7×10^9 m³ (FAO, 2014). Of the wood used for industrial processing, 40% is converted to by-products, such as sawdust, bark and black liquor, which can be subsequently burned for energy (Hakkila and Parikka, 2002). Thus, the total volume of wood used for energy each year equals 2.6×10^9 m³.

Residual forest biomass for local or regional use is typically collected over a large geographical area and transported to several end-users. Most of this material is used to produce steam, electricity and heat in industrial settings and to produce heat and electricity in towns and cities. This residual forest biomass is difficult to manage for several reasons: it is highly heterogeneous, in terms of shape and bulkiness; its distribution across the landscape is scattered and diffuse; and it is often contaminated with soil and stones, which increase the ash content at combustion, and cause equipment wear and tear (Routa et al., 2013).

Although the supply of forest biomass is generally stable, the demand for fuel varies through the year (Andersson et al., 2002). Consequently, biomass must often be stored for several months before it is used. Furthermore, biomass removal operations must be conducted with care to avoid negative impacts on soil, water and uncut trees, particularly when field conditions are sub-optimal. Forest biomass supply chains must be able to accommodate these challenges (Routa et al., 2013).

In several countries, the material used for forest biomass-based energy production consists primarily of harvesting residues generated during final fellings (see chapter: Comparison of Forest Biomass Supply Chains From the Boreal and Temperate Biomes). Thus, the residues are a by-product of roundwood harvesting. The methods used to collect harvesting residues depend on the stemwood harvesting method and the degree to which the supply chain for residue removals is integrated with roundwood removals. Whole trees are typically felled (manually or mechanically) and skidded or forwarded to the landing, where the branches and tops are removed; alternatively, trees are delimbed and topped on the cutblock, with roundwood and residues forwarded to the landing. The roundwood is then directed toward timber or pulp and paper production and the branches and tops (harvesting residues) are directed toward bioenergy production. In Nordic countries, where supply chain efficiencies for production of both roundwood and harvesting residues are probably greatest, mechanised cutting is a prerequisite for effective recovery of logging residues. Single-grip harvesters typically pile the residue in heaps on the logging site, where the material is left for a few weeks over the spring and summer to dry out. This drying period also allows the relatively nutrient-rich needles to drop from the branches, which reduces the level of nutrient removal from the site and prevents slagging and fouling of boilers during combustion. After this

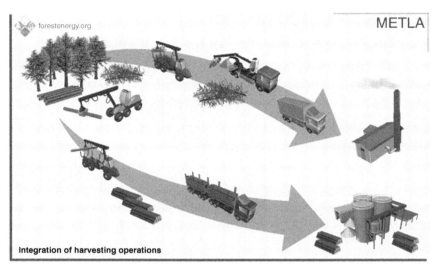

FIGURE 5.1 Integration of harvesting for roundwood and residual forest biomass. *(Source: Forest Energy Portal.)*

'seasoning' period, the residues can be loaded using a forwarder (load-carrying forest tractor) or farm tractor equipped with a grapple-loader and trailer and removed from the logging site (Fig. 5.1).

Standard forwarders can transport loose or compacted logging residues to roadside landings, where they are piled and temporarily stored, before chipping and transportation. The grapple is often modified for loose residues by removing the supporting beams that connect the ends of forks or by moving the forks inward to allow better penetration into the residue pile.

Stumps can also be harvested for bioenergy after final felling. Stumps are removed using excavators equipped with a lifting device (Laitila et al., 2008). Stumps are removed in one piece or are split into two or more pieces before being lifted. After splitting, less force is required to remove the stump from the ground and a smaller area of forest floor is disturbed. Stumps are stacked in heaps at the site for drying, which also allows rain and wind to remove soil from stump wood, improving its fuel quality. Forwarding is carried out using normal forwarders. Finland is currently the only country that uses significant amounts of stumpwood for energy (~ 1 million m^3/year).

When standing trees are harvested specifically for energy, additional felling and processing is required, compared with the collection and utilisation of harvesting residues. In Europe, the most common method currently used for forest bioenergy harvesting consists of felling and bunching trees using a single-grip harvester, or a forwarder with a modified grapple that can hold and bunch two to five trees at a time. Manual felling, using chainsaws equipped with felling handles, is also used (Heikkilä et al., 2005).

More recently, harwarders, that is combined harvester-forwarder machines, have been introduced for harvesting small trees. Harwarders fell the trees and then cut them into ~6 m lengths for forwarding. The same machines forward the material to the landing. Thus, only one machine is required to perform the whole operation. Over short forwarding distances and in small stands, harwarders are emerging as a competitive alternative to manual felling, or using harvesters with modified grapples. However, the current market share of harwarders has remained very low and, at present, harvester-forwarder systems are the most common method of cutting and extraction forest biomass for energy.

In Western Canada, non-merchantable insect-killed wood is an important source of biomass for wood pellet production (Bogdanski et al., 2011). The insect-killed wood consists largely of low-quality roundwood; therefore, the same machinery and transport methods can be used to collect high-quality roundwood for conventional forest products and low-quality insect-killed wood for wood pellet production.

Chipping of Forest Biomass

Forest biomass is typically chipped into small (10–100 mm) pieces to enable efficient handling and to improve combustion efficiency. In forests on flat terrain or in thinnings, biomass can be chipped in the forest, using an 'in-woods chipper' and then transported to the roadside. 'In-woods' chipping was popular in Nordic countries in the 1990s, but is now rarely used in Finland and Sweden because the equipment is too heavy and expensive to use for final fellings or on coarse terrain.

Most chipping is now carried out at roadside landings (Díaz-Yáñez et al., 2013). Large chippers can be mounted on trucks; smaller units can be operated using a farm tractor. The optimal size of a chipper depends on the volume of material to be processed and the condition of the forest road network. Chippers blow chips directly into trucks, which transport the chips to the plant. The main problem with these systems is that chippers cannot operate without an empty truck and trucks have to wait if the chipper has broken down. As well, the productivity of the chipper affects directly the time required to load each truck.

Forest biomass can also be chipped at a terminal. In this case, uncomminuted material is hauled to the terminal and stored. Chipping can be performed by stationary or mobile chippers. Stationary chippers are typically used in large (ie processing > 100,000 tonnes per year) terminals. Chipper productivity is higher at terminals because chips can be blown on the ground and loaded onto trucks as they become available. If the terminal is located at the end-use facility, no additional trucking is needed; chips can be fed directly into the feeding storage area using front-end loaders. If the terminal is located somewhere between the biomass source and the end-user, or if the terminal serves several end-users, then chips must be loaded into trucks and hauled. Heat or CHP plants can store chips until they are needed.

Management of the Harvesting, Chipping and Transport Processes

The transactional costs associated with the supply of forest biomass are an important determinant of the final price of the end-product (Lunnan et al., 2008). Business process mapping and re-engineering studies have illustrated the very high number (over 200) of processes involved in the purchase, harvest and delivery of a single batch of wood chips from forest to energy plant (Windisch et al., 2010, 2013). As well, the number of organisations, representatives and employees involved along the supply chain is relatively high. Forest owners, forest authorities, forest owners' associations, timber brokers and logging, chipping and transport contractors are all involved in this supply chain. In case studies conducted in Finland and Germany, ~23 person-hours were required to complete the organisational and managerial work to supply 100 m³ chips to plant (Windisch et al., 2013). This finding emphasises that increased supply chain efficiency is as dependent on developments in data and information management as it is on developments in new machinery and methods.

Predicting and Managing Fuel Quality

In response to increasing demand, actors along the supply chain have been working to ensure that the moisture content of forest biomass remains low, thereby improving its caloric value and energy density, reducing transport costs and maximising profits (Gautam et al., 2012). In the absence of artificial drying, the moisture content of solid forest biomass can vary between 50–60% (fresh wood) and 20% (Nurmi and Hillebrand, 2001; Röser et al., 2011). Thus, forest biomass can exhibit remarkable variation in total energy content per unit mass. Traditionally, studies on the natural drying process of forest biomass involved the sequential sampling and weighing of material collected from drying piles (Nurmi and Hillebrand, 2001, 2007; Röser et al., 2011). Recently, changes in the moisture content of piles of forest biomass have been monitored continuously by placing the piles on racks built on load cells (Erber et al., 2012). This methodology allows moisture changes in the entire pile of biomass to be measured, rather than just a sub-sample (Röser et al., 2011). Measurements can be taken automatically—and as frequently as required—and changing moisture patterns can be correlated with weather events, such as precipitation.

The energy content per unit mass and volume increases as forest biomass dries (Fig. 5.2). Moisture content has a direct impact on transportation efficiency, particularly when residues are so moist that the maximum weight of the load is reached before a truck has reached full capacity. Moisture contents of 50–60% are typical in freshly cut harvesting residues and small diameter energy wood; in the winter, the moisture contents of these materials can exceed 60%. With high quality storage, however, it is possible to significantly improve the energy density of forest biomass, which directly improves transportation efficiency and the sustainability of the entire supply chain.

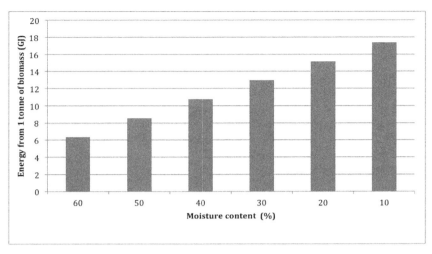

FIGURE 5.2 **Relationship between moisture content and energy content of forest biomass (note the inverse scale of the x-axis).**

If a chip truck has a net volume of 130 m³ and a maximum payload (including the weight of the fuel to run the truck) of 60,000 kg, then the average net payload of forest biomass is around 36,400 kg (Ikonen et al., 2007). If the moisture content of wood chips is over 45%, however, the net payload will be exceeded before the truck has reached its full volume (Fig. 5.3).

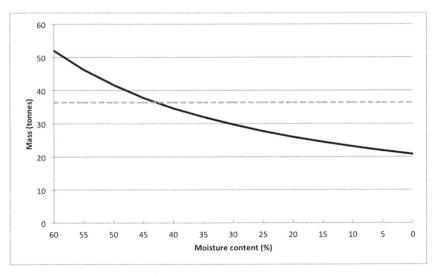

FIGURE 5.3 **The mass of 130 m³ of wood chips at different moisture contents (note the inverse scale of the x-axis).** The *dashed line* shows the maximum net payload of a typical chip truck.

The quality of forest biomass is strongly affected by changes in microclimate. In the boreal and temperate conditions of the northern hemisphere, the moisture content of wood drops rapidly in the spring. In late summer and beginning of the fall, the rate of evaporation from the surface tends to decrease and, as a consequence, the moisture content of the wood begins to rise, in some cases over and above the 'green' moisture content of freshly cut residues. Thus, it is critical to maximise the natural drying process and minimise re-moistening to preserve the fuel quality (Routa et al., 2015). The timing of harvesting operations in relation to the seasons is a crucial consideration for optimising the quality and monetary value of forest biomass.

Continuous monitoring of forest biomass as it dries can be useful, but is less meaningful if dry matter losses occur. Dry matter losses can be caused by microbial, especially fungal, activity (ie biological loss), or spillage of material during handling and storage (ie technical loss) (Pettersson and Nordfjell, 2007). Thus, the net change in mass is the sum of the water added or removed and the carbon (mainly) removed from the pile by microbial processes or spillage. The type of material, its moisture content, nutrient content and temperature, the size and shape of the biomass pile and the availability of oxygen all affect the microbial activity and degradation of forest biomass (Jirjis, 1995; Nurmi, 1999; Pettersson and Nordfjell, 2007). Needles, leaves and bark tend to be moist and nutrient-rich, encouraging fungal growth. Covering residue piles to prevent re-wetting could also suppress aeration and elevate the temperature inside the pile, thereby promoting microbial activity and accelerating dry matter loss (Jirjis, 1995; Nurmi, 1990).

During storage, dry matter losses vary between approximately 0.5 and 2.9% per month for logging residues and delimbed stems (Routa et al., 2015). These figures are supported by Jirjis (1995), who observed losses of 1.2–6.4% per month for comminuted materials. Nurmi (1999) reported losses of 0.5–1.7% per month losses for residues, while Hamelinck et al. (2005) assumed a 3% loss per month in their energy balance study. Meanwhile, Pettersson and Nordfjell (2007) reported losses of 1–1.8% per month for compacted residues.

Drying models have been developed to predict the moisture content of forest biomass based on weather conditions (Erber et al., 2012, 2014; Routa et al., 2015). There is an obvious connection between precipitation and the likelihood that the moisture content of exposed woody biomass will increase. However, other factors, such as temperature and relative humidity, also play a role (Erber et al., 2012). In Finnish conditions, Jahkonen et al. (2012) noted that the sum of potential evaporation is a good explanatory variable for the changing moisture content of logging residues. Routa et al. (2015) observed that net evaporation (the difference between evaporation and precipitation) had a strong effect on the drying process in forest biomass. Depending on the modelling approach used, wind speed, precipitation, relative humidity, evaporation and air temperature have all been found to govern the drying process; in order to ensure that these models are used appropriately, details about the climatic conditions, storage techniques and types of residues for which the models were developed must be specified (Erber et al., 2014).

Most studies examining changes in the moisture content of forest biomass have focused on the period from spring to early autumn (Nurmi, 1999; Nurmi and Hillebrand, 2007). In order to understand how the quality of this material changes through the year and what can be done to improve its heating value, it is also important to study what happens to the moisture content of woody biomass when it is piled at the roadside, where it may be stored for more than a year. Covering piles of woody residues can be used to prevent rewetting during the winter (Hillebrand and Nurmi, 2004; Röser et al., 2011), particularly in snowy climates, but, as mentioned earlier, this can also promote microbial decay.

ECONOMICS AND OVERALL QUALITY OF SUPPLY

Economic Sustainability of Feedstock Supply

Bioenergy projects must be economically viable for all actors along the forest biomass supply chain (Lunnan et al., 2008). The use of forest biomass for energy production must also be competitive with other uses for wood, for example pulp and paper. Furthermore, the energy produced from forest biomass must be cost-competitive with other energy sources. However, the costs of feedstocks and the market prices of various energy products are constantly in flux; this is particularly true of fossil fuels, which show large variations in price through time.

The availability of forest biomass, including mill residues and harvesting residues, can be extremely variable among regions. This can pose significant challenges to the financial viability of forest biomass-based energy industries, especially if much of the material is inaccessible or if its recovery rate is low. For some biomass sources, such as residues from mills, the cost of transportation is usually quite low (Welke, 2006) but for other sources, such as harvesting residues, the cost of transportation can be considerably higher and perhaps even prohibitive in some regions. This factor explains the bioenergy sector's reliance on inexpensive residues from milling operations (Tan et al., 2008). Moreover, Hughes (2000) found that, in most cases, the cost differential between biomass and coal is not sufficient to generate a profit, especially when operational and maintenance costs are included in the equation.

Nevertheless, the use of primary forest biomass, such as harvesting residues, stumpwood and whole trees from thinnings, has been cost-competitive in Nordic countries, especially for inland energy plants, where coal has become more expensive due to longer transport distances. In Finland, the number of plants using primary forest biomass for fuel has grown by two orders of magnitude since 2000 (Asikainen and Anttila, 2009). Furthermore, external costs and strategic government policies that encourage the use of forest biomass-based energy and discourage the use of fossil fuels play an important role.

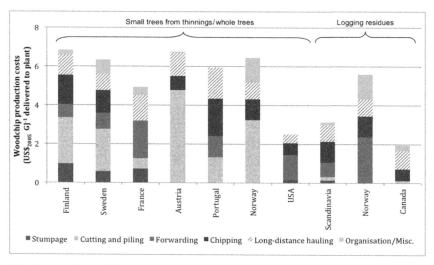

FIGURE 5.4 Cost of biomass delivery to energy plants in selected countries (Bauen et al., 2009).

Two important groups of factors affecting the costs of biomass supply are:

- The availability and quality of forest biomass within close proximity of the bioenergy plant; and
- The costs of purchasing, harvesting, processing, transporting, and storing the material.

The delivery costs for wood chips made from small trees collected during thinning operations can vary from 4 to 6.5 US$/GJ in Europe (Fig. 5.4). Chips made from harvesting residues have a lower delivery cost because they are a by-product of conventional harvesting operations for industrial roundwood: in many countries, their delivery costs is below 4 US$/GJ (Bauen et al., 2009). In British Columbia, Canada, their delivery cost was estimated to be approximately 2.7 US$/GJ (Lloyd et al., 2014); in this region, trees are mostly harvested and hauled in one piece and delimbed at roadside, making recovery of branches and tops less costly than when trees are delimbed and bucked on the clearcut area.

Sustainability can be understood in many different ways, depending on how its dimensions interact with each other and how they are emphasised in management policies and practices (Hill, 2011). In discussions of sustainability related to bioenergy production, the economic dimension tends to receive less attention than the environmental and social dimensions. Bioenergy production has many positive environmental and social impacts, such as increased employment and replacement of fossil fuels in energy production. For biomass supply chains to be sustainable, however, issues concerning economic efficiency and stability must also be considered. Economic sustainability in this context is concerned with longer-term stability. When all environmental, social and economic factors are taken into account, any activity can be considered economically sustainable if the production or activity delivers more benefits than costs over its complete life cycle (Hardisty, 2010).

Recently, increased attention has been paid to the economic sustainability and performance of the many small and medium-sized companies (SMEs) operating along forest biomass supply chains. Sustainability of the forest biomass supply chain requires that the use of forest biomass in the present-day does not compromise opportunities for future generations to use forest biomass (Lunnan et al., 2008). The main considerations are related to the long-term maintenance of soil fertility, biodiversity and watershed management; as such, they require perspectives on both forest management practices and climate change mitigation (Asikainen, 2011). Thus, while operating and pursuing profits in the short term, companies must also ensure business operations will not cause harm to the local and global environment over the long term. These kinds of commitments and guarantees cannot always be made when operating over the short-term. Decisions generally need to be made in quarter-based periods, although longer-term planning, in 2–5-year periods, are required for creating economic sustainability, local welfare and prosperity. In sustainability assessments, business operations and profitability are reviewed in 5–15-year periods (Doane and MacGillivray, 2001).

The objective of economically sustainable business practices is not only to obtain long-term profitability, but also to produce positive external effects and benefits for society and the environment. Companies operating in a sustainable way should be able to make positive contributions to their local community, broader society and the environment (Doane and MacGillivray, 2001).

According to the FAO (2010), economic sustainability of forest biomass energy has four major criteria:

- beneficial use;
- economic viability;
- economic equity; and
- property rights and landowner expectations.

Under these criteria, several indicators can be measured. These include production costs, cost-competitiveness of woody biomass as compared to competing energy resources, as well as the impacts of bioenergy production on employment rates, among others.

Economic sustainability can be analysed on a variety of scales: from individual companies to the society, or environment as a whole. Microeconomic sustainability refers to a company's ability to maintain long-term profitability (Buchholz et al., 2009). Microeconomic sustainability can be measured through a company's financial performance and its capability to manage assets. Forest biomass is often a local fuel which competes on global markets. Thus, supply chain efficiencies and costs are critical to consider when evaluating the economic sustainability of forest biomass supply.

Several internal and external factors affect the economic sustainability of the forest biomass supply chain. As mentioned earlier, the main factors determining the price of forest biomass are the costs charged to purchase the material (stumpage), the moisture content of the fuel, and the costs of harvesting and long-distance transportation to the plant or terminal. The cost structure is also dependent on energy market fluctuations, as well as labour and machine costs throughout the supply chain. In Nordic countries, the location of the comminuting phase in the whole supply chain has been shown to strongly affect the total costs of feedstock supply (Fagernäs et al., 2007). The amount of bioenergy that consumers in the market are demanding and the prices they are willing to pay define the limits of economic sustainability for the forest biomass supply chain.

Macroeconomic sustainability refers to a company's influence on the wider economy and how the company manages social and environmental impacts. Biomass and bioenergy production can improve local employment and increase tax revenues to communities (Hubbart et al., 2007). Nationally, bioenergy production and use can contribute to economic development by improving energy security and supporting local, in many cases rural, economies through employment (FAO, 2010) (see chapter: Economic and Social Barriers Affecting Forest Bioenergy Mobilisation: A Review of the Literature). On a global scale, substituting fossil fuels with biomass reduces fossil fuel consumption.

The macroeconomic sustainability of forest biomass and bioenergy, affected by the price of competing fuels, will determine the willingness of consumers to pay for biomass. International and national drivers and energy policies are used to enhance the economic sustainability of forest biomass use and bioenergy production worldwide (FAO, 2010). Prices for biomass can be affected by a variety of support mechanisms, including feed-in tariffs, subsidies and carbon taxes on fossil fuels.

In addition to positive external effects, potential external costs and adverse effects should also be considered when conducting an assessment of economic sustainability (FAO, 2010). External costs are planned or unplanned environmental and social impacts that are the result of a given process or activity, despite attempts at mitigation (Hardisty, 2010). External costs and economic sustainability are reinforced by public policies and subsidies. In fossil fuel production, external costs are particularly high, which makes renewable energy economically more sustainable. Although fossil fuels are available at a relatively low cost on the market, high CO_2 emissions, air pollution, ecological damage and other significant life-cycle external costs severely disadvantage fossil fuels, especially coal, compared with other forms of energy (Hardisty, 2010).

Overall Quality and Trust in the Supply Chain

The competitiveness of forest biomass-based energy is dependent on the cost competitiveness of the companies operating along the supply chain. Since forest bioenergy must be cost-competitive with other energy fuel sources, the cost of competing fuels defines the maximum cost for forest biomass (Ikonen et al., 2013a).

In forest biomass supply chains, it is difficult to achieve the same economies of scale as are possible with fossil fuel extraction; therefore, any additional cost efficiencies can have considerable value. Cost efficiencies can be achieved by improving the quality of biomass produced throughout the supply chain and by optimising the structure of the supply chain, which can decrease procurement costs, improve the scale of operations and make new volumes of resources available (Andersson et al., 2002). Since the cost efficiency of SMEs is not always on par with larger competitors, small bioenergy companies must focus on less tangible factors, such as product quality or customer service, in order to gain a competitive edge. Such intangible factors can have an important effect on competitiveness, especially when a company cannot compete with tangible assets (eg machinery, raw materials) or production prices.

The heating value of forest biomass is lower than that of fossil fuels and is affected by several variables, as discussed earlier. Requirements for high feedstock quality, optimal delivery times and competitive pricing are particularly strict when woody biomass is used to produce value-added energy products, such as pyrolysis oil and biodiesel.

Ultimately, final product quality will reflect the knowledge and skill of every company operating along the forest biomass supply chain. Therefore, a clear understanding of the significance of fuel quality and the ability to find ways to continually maintain or improve quality are critical. In order to ensure reliable, timely product delivery, the production of high-quality fuel also requires a good knowledge of customers' expectations and new technological advances (Röser, 2012). In addition, trust between different parties and open communication among all actors along the supply chain, from producers to consumers, is important (Huttunen, 2011).

According to Autio (2009), Pelli (2010) and Leskinen and Maier (2010), forest biomass quality is strongly determined by the availability of a skilled workforce. In an interview study from Eastern Finland, bioenergy entrepreneurs cited poor supply chain management, inadequate skills and negative attitudes as some of the biggest problems affecting forest biomass quality. These factors can also cause issues with business relationships and negatively affect a company's image as a trustworthy collaborator (Jahkonen and Ikonen, 2014). By optimising overall operational quality, SMEs can gain a competitive edge in the market.

SUMMARY AND CONCLUSIONS

Goals to mitigate climate change have brought forests and other bioenergy sources into the focus of global energy discussion. Although several countries have set targets to increase the share of renewable energy sources such as woody biomass as a climate change mitigation option (see chapter: Comparison of Forest Biomass Supply Chains From the Boreal and Temperate Biomes), under current market conditions, cost remains a major barrier to market penetration of forest bioenergy.

Wood-based fuel has been most competitive when it is a by-product of forest industry. Black liquor, bark and sawdust have had very good cost competitiveness over several decades in the European Union and also in North America. Their price at the production facility (eg sawmill yard) can even be negative, because if they are not used in energy generation, they cause waste handling costs.

Companies operating along forest biomass supply chains have to remain adaptable to both changes in the business environment and changes in supply and demand (Huttunen, 2011). Companies can reduce the risks of their operations by improving product quality and customer service (especially when it comes to deliveries). By optimising overall operational quality, SMEs can gain a competitive edge in the market. Companies that can guarantee the quality of the fuel, ensure reliable deliveries and moderate price fluctuations in the medium or long term tend to be most competitive (Ikonen et al., 2013b).

Wood has been traditionally available as a solid fuel (except black liquor and small volumes of ethanol and synthesis gas). However, market-driven replacement of oil and gas has become more promising in the liquid form. Recent technology developments enable the production of drop-in diesel for vehicles and pyrolysis oil for the replacement of heavy fuel oil. Since their market prices are much higher than that of coal, biomass-based products can penetrate the markets with modest incentives. The production of wood-based biodiesel and pyrolysis oil can be fully integrated to the production process of a modern pulp mill and large CHP plant, respectively. In this manner, all excess heat and by-products such as carbon can be fully utilised in the production of other energy products. Thus, the integration of the production of wood-based energy carriers to the forest and energy industry is the cornerstone of a market-driven replacement of fossil fuels.

REFERENCES

Andersson, G., Asikainen, A., Björheden, R., Hall, P., Hudson, J., Jirjis, R., Mead, D., Nurmi, J., Weetman, G., 2002. Production of forest energy. In: Richardson, J., Björheden, R., Hakkila, P., Lowe, A.T., Smith, C.T. (Eds.), Bioenergy From Sustainable Forestry. Kluwer Academic Publishers, Dordrecht, The Netherlands, pp. 49–123.

Asikainen, A., 2011. Raw material resources. In: Alén, R. (Ed.), Biorefining of Forest Resources: vol. 20. Papermaking Science and Technology. Paper Engineers' Association, Porvoo, Finland, pp. 115–130.

Asikainen, A., Anttila, P., 2009. Will the Growth of Forest Energy Use Continue? (in Finnish: Jatkuuko metsäenergian käytön kasvu?). Finnish Forest Research Institute, Vantaa, Finland, Available from: http://www.metla.fi/julkaisut/suhdannekatsaus/2009/suhdanne2009.pdf.

Autio, H. The skills needed in the forest energy supply chain in Middle-Finland (in Finnish). Bachelor's degree, Jyväskylä University of Applied Science (Online). Available from: http://publications.theseus.fi/bitstream/handle/10024/17630/jamk_1242194586_4.pdf?sequence=3

Bauen, A., Berndes, G., Junginger, M., Londo, M., Vuille, F., Ball , R., Bole, T., Chudziak, C., Faaij, A., Mozaffarian, H., 2009. Bioenergy—a sustainable and reliable energy source. A review of status and prospects. IEA Bioenergy (2009:06), 108 p. Available from: http://www.ieabioenergy. com/wp-content/uploads/2013/10/MAIN-REPORT-Bioenergy-a-sustainable-and-reliable-energysource.-A-review-of-status-and-prospects.pdf.

Bogdanski, B.E.C., Sun, L., Peter, B., Stennes, B., 2011. Markets for Forest Products Following a Large Disturbance: Opportunities and Challenges From the Mountain Pine Beetle Outbreak in Western Canada. Natural Resources Canada, Canadian Forest Service, Pacific Forestry Centre, Victoria, BC, Available from: https://cfs.nrcan.gc.ca/publications?id=32226.

Buchholz, T., Luzadis, V.A., Volk, T.A., 2009. Sustainability criteria for bioenergy systems: results from an expert survey. J. Clean. Prod. 17, S86–S98.

Díaz-Yáñez, O., Mola-Yudego, B., Anttila, P., Röser, D., Asikainen, A., 2013. Forest chips for energy in Europe: current procurement methods and potentials. Renew. Sustain. Energy Rev. 21, 562–571.

Doane, D., MacGillivray, A., 2001. Economic sustainability: the business of staying in business. New Economics Foundation. 52 p. Available from: http://isites.harvard.edu/fs/docs/icb.topic140232. files/RD_economic_sustain.pdf.

Erber, G., Kanzian, C., Stampfer, K., 2012. Predicting moisture content in a pine logwood pile for energy purposes. Silva Fennica 46 (4), 555–567.

Erber, G., Routa, J., Kolstrom, M., Kanzian, C., Sikanen, L., Stampfer, K., 2014. Comparing two different approaches in modeling small diameter energy wood drying in logwood piles. Croat. J. Forest Eng. 35 (1), 15–22.

Fagernäs, L., Johansson, A., Wilén, C., Sipilä, K., Mäkinen, T., Helynen, S., Daugherty, E., den Uil, H., Vehlow, J., Kaberger, T., Rogulska, M., 2007. Bioenergy in Europe: Opportunities and Barriers. VTT, Helsinki, Finland, VTT Research Note 2352. Available from: http://www.vtt.fi/ inf/pdf/tiedotteet/2006/T2352.pdf.

FAO, 2010. Criteria and Indicators for Sustainable Woodfuels. FAO Forestry Paper 160. Food and Agriculture Organization of the United Nations, Rome, Italy, Available from: http://www.fao. org/docrep/012/i1673e/i1673e00.pdf.

FAO, 2014. Global Production and Trade of Forest Products in 2013. Food and Agriculture Organization of the United Nations, Rome, Italy, Available from: http://www.fao.org/forestry/ statistics/80938/en/.

Gautam, S., Pulkki, R., Shahi, C., Leitch, M., 2012. Fuel quality changes in full tree logging residue during storage in roadside slash piles in Northwestern Ontario. Biomass Bioenergy 42, 43–50.

Hakkila, P., Parikka, M., 2002. Fuel resources from the forest. In: Richardson, J., Björheden, R., Hakkila, P., Lowe, A.T., Smith, C.T. (Eds.), Bioenergy From Sustainable Forestry. Kluwer Academic Publishers, Dordrecht, The Netherlands, pp. 19–48.

Hamelinck, C.N., Suurs, R.A., Faaij, A.P., 2005. International bioenergy transport costs and energy balance. Biomass Bioenergy 29 (2), 114–134.

Hardisty, P.E., 2010. Environmental and Economic Sustainability. CRC Press, Boca Raton, FL.

Heikkilä, J., Laitila, J., Tanttu, V., Lindblad, J., Sirén, M., Asikainen, A., Pasanen, K., Korhonen, K.T., 2005. Karsitun energiapuun korjuuvaihtoehdot ja kustannustekijät. Metlan Työraportteja 10, 8.

Hill, C., 2011. An Introduction to Sustainable Resource Use. Earthscan, London, UK; Washington, DC.

Hillebrand, K., Nurmi, J., 2004. Drying and Storage of Harvested Biomass From Young Forests (in Finnish: Nuorista metsistä korjatun energiapuun kuivatus ja varastointi). VTT Prosessit, Jyväskylä, Finland, Available from: http://jukuri.luke.fi/handle/10024/503751.

Hubbart, W.L., Biles, C., Mayfield, S.A., 2007. Sustainable Forestry for Bioenergy and Bio-Based Products. Trainers Curriculum Notebook. Southern Forest Research Partnership, Inc., Athens, GA, p. 150.

Hughes, E., 2000. Biomass cofiring: economics, policy and opportunities. Biomass Bioenergy 19 (6), 457–465.

Huttunen, M., 2011. The Current Situation of Finnish Fire Wood Market; Case: Creating a Corporate Network. Seinäjoki University of Applied Science, Seinäjoki, Finland, (in Finnish).

Ikonen, T., Asikainen, A., Prinz, R., Stupak, I., Smith, T., Röser, D., 2013a. Economic Sustainability of Biomass Feedstock Supply. IEA Bioenergy, Available from: http://ieabioenergytask43.org/wp-content/uploads/2013/09/IEA_Bioenergy_Task43_TR2013-01i.pdf.

Ikonen, T., Jahkonen, M., Pasanen, K., Tahvanainen, T., 2013b. Quality Management and Key Quality Factors of Forest Energy Supply Chains (in Finnish: Laadunhallinta ja keskeiset laatutekijät metsäenergian toimitusketjuissa). Finnish Forest Research Institute, Vantaa, Finland, Available from: http://www.metla.fi/julkaisut/workingpapers/2013/mwp275.pdf.

Ikonen, M., Palkov, A., Viljanen, K., 2007. Maximum Weight of Heavy Vehicles (in Finnish: Raskaiden ajoneuvojen omamassat). Turku University of Applied Sciences, Turku, Finland, Available from: http://www.motiva.fi/files/951/raskaiden-ajoneuvojen-omamassat-----selvitys-mahdollisuuksista-lisata-kantavuutta.pdf.

Jahkonen, M., Ikonen, T., 2014. Actors' Views on the Quality of Wood Chips Supply Chain in North Karelia (in Finnish: Toimijoiden näkemykset metsähakkeen toimitusketjun laadusta Pohjois-Karjalan alueella). Finnish Forest Research Institute, Vantaa, Finland, Available from: http://www.metla.fi/julkaisut/workingpapers/2014/mwp280.pdf.

Jahkonen, M., Lindblad, J., Sirkiä, S., Laurén, A., 2012. Predicting Energy Wood Moisture (in Finnish: Energiapuun kosteuden ennustaminen). Finnish Forest Research Institute, Vantaa, Finland, Available from: http://www.metla.fi/julkaisut/workingpapers/2012/mwp241.pdf.

Jirjis, R., 1995. Storage and drying of wood fuel. Biomass Bioenergy 9 (1), 181–190.

Laitila, J., Ranta, T., Asikainen, A., 2008. Productivity of stump harvesting for fuel. Int. J. Forest Eng. 19 (2), 37–47.

Leskinen, L., Maier, J., 2010. Forest energy strong networks (in Finnish: Metsäenergian taustalla vahvat verkostot). In: Rieppo, K. (Ed.), Growth of the Forestry and Wood Sector for Small Businesses. TTS Publications, p. 18, 406.

Lloyd, S.A., Smith, C.T., Berndes, G., 2014. Potential opportunities to utilize mountain pine beetle-killed biomass as wood pellet feedstock in British Columbia. Forestry Chron. 90 (1), 80–88.

Lunnan, A., Vilkriste, L., Wilhelmsen, G., Mizaraite, D., Asikainen, A., Röser, D., 2008. Policy and economic aspects of forest energy utilisation. In: Röser, D., Asikainen, A., Raulund-Rasmussen, K., Stupak, I. (Eds.), Sustainable Use of Forest Biomass for Energy. Springer, Dordrecht, the Netherlands, pp. 197–234.

Nurmi, J., 1990. Long-Term Storage of Fuel Chips in Large Piles (in Finnish: Polttohakkeen pitkäaikainen varastointi aumoissa). Metsantutkimuslaitos, Helsinki, Finland.

Nurmi, J., 1999. The storage of logging residue for fuel. Biomass Bioenergy 17 (1), 41–47.

Nurmi, J., Hillebrand, K., 2001. Storage alternatives affect fuelwood properties of Norway spruce logging residues. N.Z. J. Forestry Sci. 31 (3), 289–297.

Nurmi, J., Hillebrand, K., 2007. The characteristics of whole-tree fuel stocks from silvicultural cleanings and thinnings. Biomass Bioenergy 31 (6), 381–392.

Pelli, P., 2010. Biomass fuels business in central Finland. (In Finnish: Kiinteisiin biomassapolttoaineisiin liittyvä liiketoiminta Keski-Suomessa). Report 59/2010. Finnish Ministry of Employment and the Economy, Helsinki, Finland. 152 p. Available from: https://www.tem.fi/index.phtml?C=97980&s=2681&xmid=4440.

Pettersson, M., Nordfjell, T., 2007. Fuel quality changes during seasonal storage of compacted logging residues and young trees. Biomass Bioenergy 31 (11), 782–792.

Röser, D., 2012. Operational Efficiency of Forest Energy Supply Chains in Different Operational Environments. Doctorate, University of Eastern Finland (Online). Available from: http://www.metla.fi/dissertationes/df146.pdf

Röser, D., Mola-Yudego, B., Sikanen, L., Prinz, R., Gritten, D., Emer, B., Väätäinen, K., Erkkilä, A., 2011. Natural drying treatments during seasonal storage of wood for bioenergy in different European locations. Biomass Bioenergy 35 (10), 4238–4247.

Routa, J., Asikainen, A., Björheden, R., Laitila, J., Röser, D., 2013. Forest energy procurement: state of the art in Finland and Sweden. Wiley Interdisc. Rev. Energy Environ. 2 (6), 602–613.

Routa, J., Kolström, M., Ruotsalainen, J., Sikanen, L., 2015. Precision measurement of forest harvesting residue moisture change and dry matter losses by constant weight monitoring. Int. J. Forest Eng. 26 (1), 71–83.

Tan, K.T., Lee, K.T., Mohamed, A.R., 2008. Role of energy policy in renewable energy accomplishment: the case of second-generation bioethanol. Energy Policy 36 (9), 3360–3365.

Welke, S., 2006. Wood Ethanol in Canada: Production Technologies, Wood Sources and Policy Incentives. Sustainable Forest Management Network, Edmonton, Alberta, Available from: https://era.library.ualberta.ca/public/view/item/uuid:a2c342f0-dc9b-493b-9a36-66bea14525b0/.

Windisch, J., Röser, D., Sikanen, L., Routa, J., 2013. Reengineering business processes to improve an integrated industrial roundwood and energywood procurement chain. Int. J. Forest Eng. 24 (3), 233–248.

Windisch, J., Sikanen, L., Röser, D., Gritten, D., 2010. Supply chain management applications for forest fuel procurement—cost or benefit. Silva Fennica 44 (5), 845–858.

Chapter 6

Economic and Social Barriers Affecting Forest Bioenergy Mobilisation: A Review of the Literature

William A. White
Kingsmere Economics Consulting, Edmonton, AB, Canada

Highlights

- Bioenergy projects have positive impacts on key macroeconomic indicators, such as GDP and employment.
- The social impacts of bioenergy projects tend to be mixed, with worker safety and land-use change often cited as potential negative impacts.
- The use of bioenergy and other renewable energy sources can be encouraged by providing better information to consumers and producers, by improving analysis of potential social impacts before projects are put in place, in order to prevent adverse effects that will hamper social acceptance in the future and by emphasising and encouraging the potential economic and environmental benefits of bioenergy.

INTRODUCTION

Sustainable resource use requires that environmental, social and economic sustainability be demonstrated; this holds true for so-called 'green' or renewable technologies, including bioenergy. This chapter reviews the literature on the economic and social impacts of renewable resource technologies, with an emphasis on forest bioenergy.

The three legs of the sustainability stool can, at times, blend into one another. For example, a project that is not environmentally sustainable will, in all likelihood, be considered socially unacceptable and, therefore, socially unsustainable. The lines between social and economic sustainability can also be blurred. In this review, economic and social issues will be presented according to criteria described later in the chapter.

Mobilisation of Forest Bioenergy in the Boreal and Temperate Biomes

The focus of this chapter is on peer-reviewed articles appearing in scientific journals. Reports and analyses written by self-proclaimed advocates will not be included. The challenge of advocacy-based work was demonstrated by Winfield and Dolter (2014), who demonstrated that advocates of Ontario's Green Energy and Green Economy Act presented ideologically-based models that predicted significantly different outcomes than the empirical models used to assess the same policies. Since this field is rapidly changing, an emphasis will be placed on the most recent results and concepts. Papers representing a wide geographical range in both developed and developing nations will be covered. A wide range of technologies (eg, energy crops, electricity generation, etc.) and methodological approaches will be discussed. This chapter will examine not only the social and economic sustainability of forest bioenergy projects in temperate and boreal biomes, but also other forms of biomass-based energy production in other biomes. There are two main reasons for this. First, this approach allows for an examination of a broader suite of studies; with the exception of direct impacts to the forest sector, the effects of bioenergy production on the broader economy should be similar across sectors. Second, the forest sector can learn from the methodologies used by other sectors. Again, the methods used to measure the social and economic impacts of forest bioenergy projects should be similar to those employed to study bioenergy projects in other sectors (eg agriculture). That being said, the threats to sustainability can vary significantly among sectors and geographical regions. For example, forest biomass-based energy production is much less likely to result in land-use change than energy derived from agricultural products and, therefore, does not directly affect food security. Furthermore, forest crops, even fast growing ones, take a much longer time to grow than agricultural crops. Thus, agricultural and forest biomass require different levels of investment at the production stage. Once the material is harvested and used for fuel, however, these materials are treated in a relatively similar fashion.

IMPACT CLASSIFICATION AND METHODOLOGICAL APPROACHES

Economic Versus Social Impacts

Given that the field of economics is included in the social sciences, it is not surprising that there can be an overlap in what is considered an economic impact and a social impact. For example, the number of jobs created by a given energy project can be considered from an economic perspective because it results from and is a measure of economic activity. However, the creation of jobs can also be considered from a social perspective because job creation is good for society. When jobs are created in an economically distressed area, their social impact is even greater. Similarly, 'income' is clearly an economic concept, but projects that affect the distribution of wealth also have social implications. Examples of the possible economic or social impacts of bioenergy project development that will be considered in this review are given later in the chapter:

Economic

- Direct, indirect and induced employment
- GDP
- Value-added production
- Income level
- Economic welfare
- Sector viability
- Reduced costs (eg wildfire, site preparation),
- Import substitution

Social

- Land rights
- Food security
- Indoor pollution
- Food versus fuel conflicts
- Traditional practices and communities
- Worker safety
- Ethical fuel
- Gender issues

Methodological Approaches

The social and economic impacts of renewable technologies can be measured using a wide range of methodologies, many of which will be discussed here. Some papers provide overviews of several useful models while others employ a case study or scenario approach. There is no best approach to measuring social and economic impacts; methodological approaches must necessarily vary with the specific situation involved and will depend on the funding and data available.

Madlener and Myles (2000) provide an overview of several methods used to examine social and economic impacts and review a number of models that approached the problem in different ways. Of course, many more models have since been developed but this paper provides a useful overview of the many techniques available, and describes the economic and social variables that are normally considered in impact analysis. Domac et al. (2005) also provide an overview of the social and economic indices that are suitable for measuring the social and economic implications of bioenergy projects.

The most common method used to measure economic impacts is the Input–Output (I–O) model. This type of model has a strong history in economic theory and relies on data that are often readily available at the national and sub-national levels. I–O models can provide estimates of direct and indirect job creation. While useful, I–O models are not without their shortcomings. They are short-run, fixed-price models that are unidirectional (ie if one sector grows, all sectors grow) and are based on the assumption that there are sufficient unemployed workers to fulfill the requirements of newly developed projects. The models can, therefore, overstate both the economic impact of new projects and the loss of existing projects. Therefore, this methodology is frequently used for purposes of advocacy. Given that I–O models are built using tables developed for existing technologies, new technologies require additions to the tables; this can be intellectually challenging. I–O models are not useful for describing social impacts beyond employment. Nonetheless, the models can provide valuable approximations of economic impacts. The literature includes many examples of studies using I–O models. In order to demonstrate the breadth of geographical and technological applications of the methodology, these are described later in the chapter, in no particular order.

Silalertruksa and Gheewala (2010) use I–O models to investigate the impacts on employment and GDP of using cassava and molasses for bio-ethanol production in Thailand. Martínez et al. (2013) also investigate the impact of bio-ethanol production, but their study looks at sugarcane-ethanol in northeast Brazil. They study three cases: current practices (ie business as usual), more efficient agricultural and processing practices and expansion into new areas. Mukhopadhyay and Thomassin (2011) investigate the macro-economic impacts of the biofuel ethanol sector in Canada; despite the recent publication date, the study is relevant to 2003. De la Torre Ugarte et al. (2007) look at the impact across the entire United States of increasing ethanol and biodiesel to 60 billion gallons by 2030. Raneses et al. (1998) look at the economic impacts of shifting the use of cropland from food production to fuel in the United States. Kebede et al. (2013) use I–O to measure local and regional impacts of wood pellet co-firing in south and west Alabama. Perez-Verdin et al. (2008) investigate the impacts of woody biomass utilisation for bioenergy in Mississippi; they compare the use of biomass for producing biofuels or for generating electricity. Gan and Smith (2007) use east Texas to study the utilisation of logging residues for bioenergy production. Hodges et al. (2010) use I-O, among other methods, to investigate the impact of various policies and incentives for bioenergy development in Florida. Wicke et al. (2007) consider the impacts of large-scale land-use change in Argentina to expand bioenergy exports.

Social Accounting Matrix (SAM) and Computable General Equilibrium (CGE) models are rooted in the same theory as I–O models. SAM models have much the same structure as I–O models, but also have the ability to measure distributional impacts, making them more useful than I–O for measuring social impacts. Thus, SAM and CGE models can be used to determine whether a project will assist low-income earners or exacerbate income inequality (White and Patriquin, 2003). The CGE model relaxes many of the constraints of the I–O model and allows for flexible prices and multi-directional sectoral impacts. However, CGE has greater data requirements and is generally more expensive to build. The modeller must also make more decisions about parameter values and functional forms, which can lead to a stylised understanding of the economy being studied (Patriquin et al., 2007). Until recently, CGE was used for national and international level studies, but developments over the last decade have seen it used to measure sub-national impacts as well.

Kretschmer and Peterson (2010) have written a survey piece that describes existing approaches for integrating bioenergy into CGE models. They review all known multi-regional models that have employed CGE and compare their strengths and weaknesses. The authors also compare the CGE approach to the partial equilibrium approach. While each have their strengths, the authors believe CGE is more powerful, but see possibilities for combining the two approaches to achieve a better model. Hoefnagels et al. (2013) use a CGE model to estimate the impact of large-scale deployment of bioenergy for the Netherlands. Trink et al. (2010) use a regional CGE model to investigate the economic impacts of biomass-based energy services across various crops and technologies in East Styria, Austria. McDonald et al. (2006) use a CGE model to estimate the impacts of replacing some crude oil with energy from switchgrass in the United States. Hodges et al. (2010) use a combination of SAM and CGE to investigate the impact of increased use of woody biomass for energy production in Florida. The General Trade Analysis Project (GTAP) model, based on CGE modelling, is described in Kretschmer and Peterson (2010). Dandres et al. (2012) use GTAP to predict the environmental and economic impacts of European Union bioenergy policies from 2005 to 2025.

I–O and CGE models are the most popular tools for examining the economic impacts of bioenergy. However, several other methods have been employed to examine the economic and social impacts of renewable resource technologies and bioenergy in particular. An early Canadian example can be used to demonstrate the progress in impact determination and modelling: Henry et al. (1982) simply makes assumptions about the renewable energy potential of Nova Scotia's hardwood forests and then predicts the economic impacts. More recent papers, however, tend to rely on economic models.

Fang (2011) uses a Cobb–Douglas function (used by economists to measure or estimate the output from two or more inputs) to predict the economic impacts of renewable energy consumption in China. Krajnc and Domac (2007) use a spreadsheet-based model to estimate environmental and economic impacts of increased forest biomass use. While this model can be used anywhere, it was tested in Croatia and Slovenia.

Rather than developing a model to estimate impacts, some authors use existing models to run simulations, or take a scenario approach. Walsh et al. (2003) modify the partial equilibrium agricultural sector model POLYSYS to estimate the economic impact of bioenergy crop production in the United States. Similarly, Schwarzbauer and Stern (2010) use a simulation approach to look at the economic impacts of a wood for energy scenario on Austria's forest sector.

There are also studies that focus on providing stakeholders with tools with which to measure or estimate social and economic impacts. This type of study usually includes the development of indicators, or a method to collect indicators that can be replicated by interested parties. For example, Hoffmann (2009) suggests a list of variables and parties to be interviewed to collect information that would allow regional authorities to estimate the impacts of various bioenergy utilisation technologies. The general nature and application of this approach is demonstrated by the fact that no particular country is used in the description within the paper. Dale et al. (2013) also use an indicator approach to assess the economic and social sustainability of bioenergy systems in Finland. Ayoub et al. (2007) outline a system that allows project planners to collect their own information for estimating the impacts of bioenergy production. They demonstrate a Decision Support System (DSS) that allows planners to consider every stakeholder along the biomass supply chain, their possible social concerns, as well as their potential economic and environmental impacts.

Faaij and Domac (2006) provide an overview of the potential for bioenergy markets and look at possible economic impacts and barriers to development. No case studies are included, but this paper provides a useful overview for non-economists.

The techniques and methodologies described above focus primarily on ways to estimate economic impacts. The methods outlined in the remainder of this section are particularly useful for determining social impacts. Most of the papers discussed below involve smaller regions and more direct interactions with stakeholders, through interviews or surveys; these types of data are more difficult to collect on a larger scale. The first example involves two case studies conducted in Tanzania by Van Eijck et al. (2014a). They investigate the social, economic and environmental impacts of both smallholder and plantation-based jatropha production systems. By focusing on small areas, the authors are able to collect their own data and use secondary sources. Van Dam et al. (2009) study regional impacts of bioenergy production from soya beans and switchgrass in Argentina, using a case-study approach.

Surveys are another means of collecting detailed data from small areas. Laramee and Davis (2013) present information related to domestic bio-digesters from forty households in Tanzania. This study reports information on economic and environmental factors, as well as social impacts, such as reduced time to procure energy. He et al. (2013) also use a survey method to compare centralised and decentralised bioenergy systems in rural China, with a focus on social impacts. Haughton et al. (2009) use direct stakeholder contact to identify key indicators for recognising success and assessing the social and economic impacts of planting perennial biomass crops in the United Kingdom. Stakeholders are also the key respondents in a study by Obidzinski et al. (2012) of palm oil plantations and biofuel production in Indonesia. Again, by focusing on a small number of stakeholders, the authors provide more information on social impacts than can be obtained from data-rich models. Although surveys can be used to collect short answers to a large number of questions, interviews can provide a large amount of information from fewer respondents. Again, this is important in ascertaining social impacts, where qualitative data can be of paramount importance. Rossi and Hinrichs (2011) present the results of interviews with farmers and local community members in southern Iowa and north-eastern Kentucky to assess the local impacts of agricultural bioenergy. Hoffmann (2009) also uses interviews to obtain data for model development, demonstrating that interviews are not only useful for collecting qualitative data. Mapping exercises can also be used by researchers to determine the areas that are most sensitive to social impacts. Haughton et al. (2009) in the United Kingdom and Phalan (2009) in Asia provide examples of this.

Finally, multi-criteria analysis (MCA) attempts to look simultaneously at social, economic and ecological (or environmental) impacts of bioenergy production. MCA uses formal approaches for utilising multiple criteria, in order to help groups or individuals make decisions. The multiple criteria can be in the form of overarching social, economic or ecological goals, or can be used to bring together diverse viewpoints from different stakeholders. Buchholz et al. (2007) compare the results from four MCA tools employed in a multi-stakeholder bioenergy case study in Uganda. Elghali et al. (2007) use multi-criteria decision analysis (MCDA, similar to MCA) to integrate and reconcile the interests and concerns of diverse stakeholder groups. The two examples cited here demonstrate the diverse applications of the approach.

There is a wide range of techniques available to measure or estimate the social and economic impacts of bioenergy projects. These include the use of existing data and models, the collection of primary data, the development of new models, or combinations of these. Data can take the form of quantitative measurements, such as employment and GDP, or interviews used to assess less tangible variables, such as wellbeing or land rights. There is no one correct model or methodological approach; the method used to evaluate social and economic impacts will depend on what the author or proponent of the study wishes to determine, the availability of data and the budget and resources available.

REVIEW OF RESULTS

The economic and social impacts of bioenergy projects are reported in separate sections below. Where a paper considers both economic and social sustainability, the results will be divided between these sections. The papers will be grouped by the type of bioenergy produced or the sector studied, if no specific form of bioenergy is discussed.

Economic Impacts

Silalertruksa and Gheewala (2010) found that bio-ethanol production in Thailand increased GDP and required 17–20 more workers than petrol, for the same amount of energy. Most (90%) of this additional employment was in the agricultural sector. Although bio-ethanol production required greater imports, these were partially offset by decreased petrol imports. Van Eijck et al. (2014a) compared the economic impacts of producing biofuel from jatropha grown in plantations or in small-holder cultivation systems. The plantation model created greater employment and higher local prosperity than the smallholder model, although both had positive impacts. These benefits were put at risk by the marginal profitability of production. Ohimain (2012) found that replacing fuelwood and kerosene with cassava-derived bio-ethanol created jobs and boosted rural agriculture in Nigeria. Martínez et al. (2013) considered three scenarios for sugarcane-ethanol production in Brazil: business as usual, more efficient agricultural and processing practices, or expansion into new areas. All scenarios resulted in increased value-added production in the region. Although employment directly associated with production decreased because of mechanisation, total employment in the region increased. Many of the additional jobs were expected in regions outside of sugarcane production. In fact, Azadi et al. (2012) and Mukhopadhyay and Thomassin (2011) also concluded that increased bio-ethanol production would raise employment levels in Brazil and Canada, respectively. Biofuel production from woody biomass was shown to have greater positive effects on the Mississippi economy than the production of electricity, due to differences in operating and construction costs (Perez-Verdin et al., 2008), while De la Torre Ugarte et al. (2007) found that a significant expansion in ethanol and biodiesel production in the United States would impact the economy positively, particularly the agricultural sector. Demirbas (2009) provided an overview of the economic implications of increased biofuel production and listed the following impacts: greater fuel diversity, improved employment, higher government revenues, increased investment in plant and equipment, decreased dependence on imported petroleum and enhanced agricultural development.

Many bioenergy projects use feedstock directly sourced from the forest, or residues from the forest industry. Caurla et al. (2013) examined the impact of increasing fuelwood consumption in France via public policies. They did not discuss employment impacts, but noted positive effects for both the forest sector and consumers. Polagye et al. (2007) conducted an analysis of increased bioenergy production using thinnings. Although the use of thinnings for bioenergy (co-firing) does not completely cover the cost of the thinning operations, it reduces the risk of wildfire in overstocked forests. Kebede et al. (2013) studied the economic impact of using wood pellets for co-firing in south and west Alabama and found positive economic impacts in the form of increased employment, income and value-added production, as well as substitution of imported coal for local wood resources. Pa et al. (2013) examined the effect of substituting fuelwood with wood pellets in the Canadian province of British Columbia. The authors found significant savings in fuel costs and improvements in human health, largely resulting from decreased emissions. Hodges et al. (2010) reported that increased use of woody biomass in Florida would increase GDP, raise employment and enhance government revenues. Although the forest sector would see economic gains, the manufacturing sector would see declines resulting from increased feedstock prices. In their study of wood-for-energy scenarios, Schwarzbauer and Stern (2010) reported similar results for the Austrian paper and panel industry. Forest companies and sawmills benefited from higher prices for logs and residues; these same cost increases hurt the paper and panel industries. Schwarzbauer et al. (2013) also looked at the impact of economic crises (local, export and global) on the forest sector and found that fuelwood prices and production both declined. Trink et al. (2010) showed that there were positive employment effects for all forest biomass-based energy projects studied in East Styria, Austria. Manley and Richardson (1995) reported positive economic benefits from wood energy in Sweden, Canada and Switzerland, although few numerical details were provided. Gan and Smith (2007) reported that the use of logging residues for bioenergy production in east Texas increased employment and value-added production and reduced the cost of site preparation. Obidzinski et al. (2012) reported significant economic gains for investors, farmers and employees from palm oil plantations in Indonesia but, as described in the section on social impacts, these did not come without social costs.

Although forests are an important source of biomass for energy, the literature revealed many studies that highlight the economic impacts of bioenergy from agriculture. Trink et al. (2010) showed that the increased land intensity of agricultural biomass production crowds out conventional agriculture and, in many cases, results in reduced profits, higher land prices and increased costs to the consumer. Rossi and Hinrichs (2011) reported that farmers interviewed in north-eastern Kentucky and southern Iowa were hopeful that bioenergy projects would revitalise the area, but were concerned that profits would benefit only non-local corporate enterprises. Walsh et al. (2003) demonstrated positive economic impacts for the US agricultural sector from bioenergy crop production. In a rare negative example, however, McDonald et al. (2006) used switchgrass as a case study to demonstrate that energy efficiency is reduced and GDP drops in the United States when biofuel replaces crude oil. Raneses et al. (1998) found that some sectors fare better than others, when cropland is shifted from food to switchgrass (for fuel) production. For example, consumers will pay higher prices for food and livestock because producers face higher feed costs, while switchgrass producers benefit. Greek farmers achieved increased profits and higher employment, when energy crops were grown for heat production (Panoutsou, 2007). A study of 40 households in Tanzania suggested that farm incomes increased and energy-related expenditures decreased when domestic bio-digesters were installed (Laramee and Davis, 2013). He et al. (2013) found that centralised and household digesters were both found to improve incomes for local farmers and villagers in China.

A number of studies have also been conducted to assess the effects of bioenergy project development at the macro-economic scale. Positive employment impacts were reported by Hoefnagels et al. (2013), who looked at the effect of large-scale deployment of biomass resources for energy in the Netherlands. Weldegiorgis and Franks (2014) compared coal and biomass as energy sources in Australia and found that biomass suppliers contributed significantly to direct employment at the regional level. Large-scale land-use change and export-oriented bioenergy production in Argentina was investigated by Wicke et al. (2007), who reported very positive impacts on trade balance, GDP and employment. Similarly, Van den Broek et al. (2000) found that electrical generation from eucalyptus and bagasse by sugar mills in Nicaragua resulted in twice the employment of that from conventional fuel and a quadrupling of GDP. In an example from the United Kingdom, Thornley et al. (2008) showed that power-only biomass systems created 1.27 person years of employment per GWh, while CHP systems created more than 2 person years per GWh.

Social Impacts

The following paragraphs are organised by project type and include a discussion of biofuels in general, forest biomass-based energy, bioenergy in general and bio-digesters.

Van Eijck et al. (2014a) reported that small-holder biofuel production from jatropha was more successful at preserving land rights and biodiversity than plantation production. Ohimain (2012) reported that bio-ethanol from cassava reduced indoor pollution, compared with the use of fuelwood and kerosene, although there was increased potential for food versus fuel conflicts. Sawyer (2008) pointed out the risks associated with increased biofuel production at the Brazilian frontier. These include: worker safety, conversion of forests to pasture and the disruption of family farms and traditional communities. Van der Horst and Vermeylen (2011) investigated the social impacts of biofuels across a variety of geographical regions. They stated that the social impacts of biofuel production in developed countries would be relatively minor and could be mitigated by progressive social policy. Positive social impacts of bioenergy projects in developing countries are possible but require government intervention to ensure that the poor receive a fair share of the benefits; supply chain certification could be employed to ensure that biofuel is produced ethically. Obidzinski et al. (2012) found that palm oil plantations in Indonesia were profitable but did not distribute their benefits equally. As well, stakeholders not directly involved in the plantation, for example traditional landowners, were vulnerable to greater restrictions on traditional land-use rights. Land scarcity, rising land prices and conflicts over land-use were also common when plantations were present. Hazelton et al. (2013) focused on the social impacts of eight modern forest-based bioenergy feedstock production models in Uganda. Small private production facilities could be profitable but provided few benefits to the landless poor. Larger projects could produce greater financial benefits but the natural resource impacts could harm neighbouring communities. Bioenergy initiatives that allowed the landless poor to have a collaborative stake were the most successful in achieving rural development objectives.

Farmers and local community members were interviewed by Rossi and Hinrichs (2011) to determine their views on the benefits of bioenergy projects in southern Iowa and north-eastern Kentucky. Respondents were sceptical of the social impacts of these projects because of their expectation that corporate interests would over-ride farmer's priorities. Van Eijck et al. (2014b) conducted an extensive global review of experience with jatropha but found insufficient information to reach definitive conclusions with respect to social impacts. As mentioned earlier, Weldegiorgis and Franks (2014) reported higher rates of workplace injuries with bioenergy alternatives compared with the use of fossil fuels. They also noted significant changes in land-use in some scenarios.

Laramee and Davis (2013) found that a program of domestic bio-digesters in Tanzania could reduce the time required to procure energy, a task that normally falls to women. He et al. (2013) surveyed officials, village leaders and farmers in Shandong province, China, to compare the benefits of centralised versus decentralised (in-home) bio-digesters. Both were found to have positive social impacts, but the centralised model was more favourable in terms of household workload and improved sanitation.

BARRIERS TO BIOENERGY MARKET DEVELOPMENT

This section includes a discussion of existing and potential barriers to bioenergy development across the world. Barriers will be looked at from the point of view of consumers, firms or producers and the government. Much of this discussion is drawn from White et al. (2007).

Consumers have a variety of choices with respect to using bioenergy in the developed world. A consumer may choose a furnace, stove or boiler that uses pellets, chips, or other woody material for residential space heating and/or hot water. A district heating plant powered by bioenergy may provide service in their area. In some cases, bioenergy-based electricity may also be available. Questions, however, remain: what will lead households to increase their consumption of bioenergy and what barriers exist to discourage households from choosing bioenergy?

For many households, the costs of investment in bioenergy infrastructure and of bioenergy production are the primary concern. A secondary consideration is the security and comfort of bioenergy supply, compared with other forms of energy. Will there be enough wood pellets in the market? Is it just 'the environmentalists' flavour of the month', or will bioenergy be easily available in the market for an indefinite period? For the market system to operate efficiently, therefore, consumers must have sufficient knowledge about the goods and services available, their price and the benefits to be gained from their use. Economic theory assumes that information on costs, environmental effects, the quality and reliability of service and the quality of heat (eg heat from a district heating plant or wood pellet stove compared with a natural gas furnace) are understood by the household before a decision is made. Clearly, this assumption rarely, if ever, applies in the real world and, as a result, the allocation of resources towards bioenergy production is never optimal. Information can be used as a tool used by industry (private marketing), non-governmental organisations (social marketing) and the government in an attempt to modify consumer behaviour (Cash et al., 2006). Consumers must be aware that information exists and know where to look for it (Clark, 2004).

The other dimension of this issue is the 'information paradox', in which a consumer has too much information and cannot assess the important points by her/himself. Liberalisation of the energy market and bioenergy itself remain a novelty in some countries. The lack of accurate information can make it very difficult for households to make optimal choices with respect to new energy technologies. Capital investments are usually among the most expensive purchases a household undertakes and, as a result, caution and diligence are exercised. Concerns about individual heat control, furnace maintenance, fire hazards, etc., require time and effort to address. This challenge is compounded when consumers are sceptical of the accuracy of information they may receive from private companies (this holds true for both conventional energy suppliers and sellers of new technologies) and governments. Households make choices about information gathering in the same way they choose any other good or service; they evaluate the costs and benefits of collecting that information and choose an optimal amount. In the case of bioenergy, most households have too little information; as a result, the cost of assessing and/or collecting information is particularly high and leads to a slow uptake of new technology.

A third consideration for many people can be the perceived effect on the community and/or the environment. If the bioenergy supply is of local origin, it may provide employment opportunities and/or improved social welfare, due to an expanded tax base. On the other hand, locally supplied bioenergy may be perceived by some as harmful to the environment if they believe forests are being cut unsustainably or air quality is impaired. If faced with tight budget constraints, the desire to be 'a good environmental citizen' may be perceived as less important than having available funds to purchase other necessary goods and services.

Profitability, either from cost-savings or increased/sustained market share, is behind any economic driver. Ikonen et al. (2013) provide a useful discussion of the economic sustainability of bioenergy (see also chapter: Challenges and Opportunities of Logistics and Economics of Forest Biomass). Bioenergy companies can become agents in the bioenergy market, in response to pure market forces or in response to government regulations and/or subsidies and taxes. Given that profitability is the underlying driver that encourages firms to invest in bioenergy (for either cost-saving or income-generating motives), it falls to governments to foster long-term market development in order to provide an environment that ensures long-term profitability. Although short-term losses may be acceptable if sustainable profits are expected in the future, an individual company cannot realistically enter a market without the hope of sustainable profits in the long-term. Profitability may only be possible with targeted government support. However, companies are unlikely to enter the market if government assistance is not expected to continue or if the firm does not foresee profitability without subsidies. Unclear and time-consuming procedures for obtaining project approval can stymie investment in bioenergy projects and direct government intervention can reduce profits, or even drive firms from their core businesses (eg via abatement costs, polluter-pays principles, etc.). The lack of existing supply chains and services may also discourage firms from entering the bioenergy market (Silveira et al., 2006). In addition, a lack of support from the financial sector, in terms of loans or high investment costs, can be an important obstacle.

The lack of capital also represents an important obstacle for individual companies. Bioenergy utilisation necessitates a significant level of investment that might require support from financial institutions. Other drivers that may discourage the uptake of bioenergy by private companies are not dissimilar to those that face households. Lack of information and know-how, concerns about the long-term supply of feedstock, perceived risks associated with change, and concerns about access to services could all impede investment in bioenergy. Bioenergy firms looking to enter the heat, electricity and motor fuel markets must compete with firms whose structures are already highly centralised, have significant market power and who benefit from creating barriers to new competitors. At the same time, new companies can be at a cost disadvantage compared with the economies of scale accessible to established firms and can be prevented from performing 'hit and run' strategies, due to economies of scope (Baumol et al., 1982). The energy sector is typically characterised by significant sunk costs (eg regulatory approvals, etc.) that can represent an important financial barrier (Fulton and Giannakas, 2001). More established firms can also have influence over prices and access to information, which make it difficult for consumers to choose new energy sources. In such an environment, economic theory on market structure indicates that bioenergy cannot compete in a sustainable way with conventional sources of energy without some kind of targeted government support. If, however, smaller bioenergy production plants were isolated from the existing energy markets with which they currently compete, it is possible to imagine a decentralised energy system. In such a system, with many small buyers and sellers that are too small to influence the overall price of energy delivered, the market might develop the characteristics of perfect competition.

By liberalising energy markets, the influence of large firms facing little or no competition was expected to be reduced. Instead, publicly owned energy companies have been converted into privately-owned monopolies at the national level. This is especially true for most Central and Eastern European countries (eg Poland, Croatia, Romania, Hungary). Elsewhere, traditionally market-driven economies have developed into oligopolies (eg the United Kingdom), particularly in the electricity and fuel markets (Clark, 2007; Toke, 2007). Market liberalisation in Canada has also led to widespread dissatisfaction; the new market structure has been blamed, rightly or wrongly, for higher electricity and heating prices (Dewees, 2005). Wholesale electricity restructuring in the United States has not developed sufficient wholesale competition and the 'power to choose' energy suppliers has, in many cases, not reduced costs for residential and small commercial electricity consumers (Kelly and Moody, 2005).

In some countries, targeted government policies have been successful in encouraging bioenergy consumption and removing some of the barriers to bioenergy production discussed earlier. If a government is convinced renewable energy is beneficial, it must implement policies to encourage investment by producers and consumers. These policies can include establishment of subsidies or feed-in tariffs, provision of timely and accurate information, facilitation of free and fair trade, formulation of consistent and reliable guidance, as well as development of clear funding objectives and credible research programs. While there are many barriers to the implementation of bioenergy and other renewable energy sources, the increasing importance of 'green' energy demonstrates that barriers can be overcome. Indeed, McCormick and Kåberger (2007) state that consistent government strategies and policy interventions can overcome existing barriers.

SUMMARY AND CONCLUSIONS

A wide body of literature exists examining the social and economic impacts of renewable energy sources, and bioenergy in particular. Many methodologies, ranging from modelling exercises to individual interviews, have been shown to provide critical information about the social and economic impacts of bioenergy projects in a variety of settings across the globe and over a range of biomass feedstocks. Employment and GDP are the focus of most economic impact studies, while time savings in fuel collection, rural employment, land-use change, impact on traditional communities and food versus fuel debates are key social impacts. The literature is overwhelmingly positive in its assessment of the economic impacts of bioenergy. That is, most projects reviewed resulted in positive changes in the economic indicators outlined in the beginning of the chapter. A negative social impact noted in a few papers was the increase in worker injuries, especially during forest-based biomass procurement. Issues such as food versus fuel and indirect land-use change have been identified as possible negative social impacts of bioenergy development, but these mainly apply to new agricultural or forestry plantations and not to the sustainable use of existing forests.

There are far fewer studies published on the social impacts of bioenergy production than on economic impacts and most of these studies were undertaken in the developing world. In fact, not a single study was located that quantifies the social impact of forest biomass-based energy systems in temperate or boreal forests. Given that forest biomass constitutes a major fraction of global biomass potential, there is a clear need for more work on the economic and, especially, the social impacts of forest biomass-based energy production in the boreal and temperate forest regions. Nevertheless, the existing literature provides useful insights into the potential opportunities and barriers to the mobilisation of supply chains in the forest context. As more forest biomass-based energy projects are put in place, greater opportunities will be available to study their impact on individuals, communities, regions and countries. Appropriate methodologies will be further refined and, in time, predictive models will be developed to assist developers and governments in overcoming existing barriers for new projects, over a variety of scales and locations and in understanding the expected economic and social impacts.

That being said, the existing literature suggests that the use of bioenergy and other renewable energy sources can be encouraged by:

- Providing better information to consumers and producers;
- Improving analysis of potential social impacts before projects are put in place, in order to prevent adverse effects that will hamper social acceptance in the future; and
- Emphasising and encouraging the potential economic and environmental benefits of bioenergy.

REFERENCES

Ayoub, N., Martins, R., Wang, K., Seki, H., Naka, Y., 2007. Two levels decision system for efficient planning and implementation of bioenergy production. Energy Conv. Manag. 48 (3), 709–723.

Azadi, H., de Jong, S., Derudder, B., De Maeyer, P., Witlox, F., 2012. Bitter sweet: how sustainable is bio-ethanol production in Brazil? Renew. Sustain. Energy Rev. 16 (6), 3599–3603.

Baumol, W.J., Panzar, J.C., Willig, R.D., Bailey, E.E., Fischer, D., 1982. Contestable Markets and the Theory of Industry Structure. Harcourt Brace Jovanovich, New York.

Buchholz, T.S., Volk, T.A., Luzadis, V.A., 2007. A participatory systems approach to modeling social, economic, and ecological components of bioenergy. Energy Policy 35 (12), 6084–6094.

Cash, S.B., Goddard, E.W., Lerohl, M., 2006. Canadian health and food: the links between policy, consumers, and industry. Can. J. Agric. Econ. 54 (4), 605–629.

Caurla, S., Delacote, P., Lecocq, F., Barkaoui, A., 2013. Stimulating fuelwood consumption through public policies: an assessment of economic and resource impacts based on the French forest sector model. Energy Policy 63, 338–347.

Clark, W.W., 2004. Forget about liquefied natural gas: we need diverse clean energy now. Electric. J. 17 (8), 87–90.

Clark, W.W., 2007. Partnerships in creating agile sustainable development communities. J. Clean. Prod. 15 (3), 294–302.

Dale, V.H., Efroymson, R.A., Kline, K.L., Langholtz, M.H., Leiby, P.N., Oladosu, G.A., Davis, M.R., Downing, M.E., Hilliard, M.R., 2013. Indicators for assessing socioeconomic sustainability of bioenergy systems: a short list of practical measures. Ecol. Indicators 26, 87–102.

Dandres, T., Gaudreault, C., Tirado-Seco, P., Samson, R., 2012. Macroanalysis of the economic and environmental impacts of a 2005–2025 European Union bioenergy policy using the GTAP model and life cycle assessment. Renew. Sustain. Energy Rev. 16 (2), 1180–1192.

De la Torre Ugarte, D., English, B.C., Jensen, K.L., 2007. Sixty billion gallons by 2030 economic and agricultural impacts of ethanol and biodiesel expansion. Paper presented in the 2007 Annual Meeting, July 29–August 1, 2007, Portland, Oregon TN: American Agricultural Economics Association (new name 2008: Agricultural and Applied Economics Association).

Demirbas, A., 2009. Political, economic and environmental impacts of biofuels: a review. Appl. Energy 86 (Suppl. 1), S108–S117.

Dewees, D.N., 2005. Electricity restructuring and regulation in the provinces: Ontario and beyond. Presented to Energy, Sustainability and Integration—The CCGES Transatlantic Energy Conference, York University, Toronto, Canada, 9–10 September 2005.

Domac, J., Richards, K., Risovic, S., 2005. Socio-economic drivers in implementing bioenergy projects. Biomass Bioenergy 28 (2), 97–106.

Elghali, L., Clift, R., Sinclair, P., Panoutsou, C., Bauen, A., 2007. Developing a sustainability framework for the assessment of bioenergy systems. Energy Policy 35 (12), 6075–6083.

Faaij, A.P., Domac, J., 2006. Emerging international bio-energy markets and opportunities for socio-economic development. Energy Sustain. Dev. 10 (1), 7–19.

Fang, Y., 2011. Economic welfare impacts from renewable energy consumption: the China experience. Renew. Sustain. Energy Rev. 15 (9), 5120–5128.

Fulton, M., Giannakas, K., 2001. Agricultural biotechnology and industry structure. AgBioForum 4 (2), 137–151.

Gan, J., Smith, C., 2007. Co-benefits of utilizing logging residues for bioenergy production: the case for East Texas, USA. Biomass Bioenergy 31 (9), 623–630.

Haughton, A.J., Bond, A.J., Lovett, A.A., Dockerty, T., Sünnenberg, G., Clark, S.J., Bohan, D.A., Sage, R.B., Mallott, M.D., Mallott, V.E., Cunningham, M.D., Riche, A.B., Shield, I.F., Finch, J.W., Turner, M.M., Karp, A., 2009. A novel, integrated approach to assessing social, economic and environmental implications of changing rural land-use: a case study of perennial biomass crops. J. Appl. Ecol. 46 (2), 315–322.

Hazelton, J.A., Windhorst, K., Amezaga, J.M., 2013. Forest based biomass for energy in Uganda: stakeholder dynamics in feedstock production. Biomass Bioenergy 59, 100–115.

He, G., Bluemling, B., Mol, A.P., Zhang, L., Lu, Y., 2013. Comparing centralized and decentralized bio-energy systems in rural China. Energy Policy 63, 34–43.

Henry, G., Hanson, A., Freedman, B., 1982. The renewable energy potential of Nova Scotia's hardwoods. Biomass 2 (2), 139–151.

Hodges, A.W., Stevens, T.J., Rahmani, M., 2010. Economic Impacts of Expanded Woody Biomass Utilization on the Bioenergy and Forest Products Industries in Florida. University of Florida, Institute of Food and Agricultural Sciences, Food and Resource Economics Department, Gainesville, FL, Available from: https://www.gru.com/Portals/0/Legacy/Pdf/futurePower/TestimonialsExhibits/Schroeder/RMS-7DivofForestry2-23-10.pdf.

Hoefnagels, R., Banse, M., Dornburg, V., Faaij, A., 2013. Macro-economic impact of large-scale deployment of biomass resources for energy and materials on a national level—a combined approach for the Netherlands. Energy Policy 59, 727–744.

Hoffmann, D., 2009. Creation of regional added value by regional bioenergy resources. Renew. Sustain. Energy Rev. 13 (9), 2419–2429.

Ikonen, T., Asikainen, A., Prinz, R., Stupak, I., Smith, T. and Röser, D., 2013 Economic Sustainability of Biomass Feedstock Supply. Available from: http://ieabioenergytask43.org/wp-content/uploads/2013/09/IEA_Bioenergy_Task43_TR2013-01i.pdf

Kebede, E., Ojumu, G., Adozssi, E., 2013. Economic impact of wood pellet co-firing in South and West Alabama. Energy Sustain. Dev. 17 (3), 252–256.

Kelly, S., Moody, D., 2005. Wholesale electric restructuring: was 2004 the "Tipping Point"? Electric. J. 18 (2), 11–18.

Krajnc, N., Domac, J., 2007. How to model different socio-economic and environmental aspects of biomass utilisation: case study in selected regions in Slovenia and Croatia. Energy Policy 35 (12), 6010–6020.

Kretschmer, B., Peterson, S., 2010. Integrating bioenergy into computable general equilibrium models—a survey. Energy Econ. 32 (3), 673–686.

Laramee, J., Davis, J., 2013. Economic and environmental impacts of domestic bio-digesters: evidence from Arusha, Tanzania. Energy Sustain. Dev. 17 (3), 296–304.

Madlener, R., Myles, H., 2000. Modelling Socio-Economic Aspects of Bioenergy Systems: A Survey Prepared for IEA Bioenergy Task 29. Available from: http://www.task29.net/assets/files/reports/Madlener_Myles.pdf

Manley, A., Richardson, J., 1995. Silviculture and economic benefits of producing wood energy from conventional forestry systems and measures to mitigate negative impacts. Biomass Bioenergy 9 (1), 89–105.

Martínez, S.H., van Eijck, J., da Cunha, M.P., Guilhoto, J.J., Walter, A., Faaij, A., 2013. Analysis of socio-economic impacts of sustainable sugarcane–ethanol production by means of interregional Input–Output analysis: demonstrated for Northeast Brazil. Renew. Sustain. Energy Rev. 28, 290–316.

McCormick, K., Kåberger, T., 2007. Key barriers for bioenergy in Europe: economic conditions, know-how and institutional capacity, and supply chain co-ordination. Biomass Bioenergy 31 (7), 443–452.

McDonald, S., Robinson, S., Thierfelder, K., 2006. Impact of switching production to bioenergy crops: the switchgrass example. Energy Econ. 28 (2), 243–265.

Mukhopadhyay, K., Thomassin, P.J., 2011. Macroeconomic effects of the ethanol biofuel sector in Canada. Biomass Bioenergy 35 (7), 2822–2838.

Obidzinski, K., Andriani, R., Komanidin, H., Andrianto, A., 2012. Environmental and social impacts of oil palm plantations and their implications for biofuel production in Indonesia. Ecol. Soc. 17 (1), 25.

Ohimain, E.I., 2012. The benefits and potential impacts of household cooking fuel substitution with bio-ethanol produced from cassava feedstock in Nigeria. Energy Sustain. Dev. 16 (3), 352–362.

Pa, A., Bi, X., Sokhansanj, S., 2013. Evaluation of wood pellet application for residential heating in British Columbia based on a streamlined life cycle analysis. Biomass Bioenergy 49, 109–122.

Panoutsou, C., 2007. Socio-economic impacts of energy crops for heat generation in Northern Greece. Energy Policy 35 (12), 6046–6059.

Patriquin, M.N., Wellstead, A.M., White, W.A., 2007. Beetles, trees, and people: regional economic impact sensitivity and policy considerations related to the mountain pine beetle infestation in British Columbia, Canada. Forest Policy Econ. 9 (8), 938–946.

Perez-Verdin, G., Grebner, D.L., Munn, I.A., Sun, C., Grado, S.C., 2008. Economic impacts of woody biomass utilization for bioenergy in Mississippi. Forest Prod. J. 58 (11), 75–83.

Phalan, B., 2009. The social and environmental impacts of biofuels in Asia: an overview. Appl. Energy 86 (Suppl. 1), S21–S29.

Polagye, B.L., Hodgson, K.T., Malte, P.C., 2007. An economic analysis of bio-energy options using thinnings from overstocked forests. Biomass Bioenergy 31 (2), 105–125.

Raneses, A., Hanson, K., Shapouri, H., 1998. Economic impacts from shifting cropland use from food to fuel. Biomass Bioenergy 15 (6), 417–422.

Rossi, A.M., Hinrichs, C.C., 2011. Hope and skepticism: farmer and local community views on the socio-economic benefits of agricultural bioenergy. Biomass Bioenergy 35 (4), 1418–1428.

Sawyer, D., 2008. Climate change, biofuels and eco-social impacts in the Brazilian Amazon and Cerrado. Philos. Trans. R. Soc. B Biol. Sci. 363 (1498), 1747–1752.

Schwarzbauer, P., Stern, T., 2010. Energy vs. material: economic impacts of a "wood-for-energy scenario" on the forest-based sector in Austria—a simulation approach. Forest Policy Econ. 12 (1), 31–38.

Schwarzbauer, P., Weinfurter, S., Stern, T., Koch, S., 2013. Economic crises: impacts on the forest-based sector and wood-based energy use in Austria. Forest Policy Econ. 27, 13–22.

Silalertruksa, T., Gheewala, S.H., 2010. Security of feedstocks supply for future bio-ethanol production in Thailand. Energy Policy 38 (11), 7476–7486.

Silveira, S., Andersson, L., Lebedys, A., 2006. Opportunities to boost bioenergy in Lithuania. Biomass Bioenergy 30 (12), 1076–1081.

Thornley, P., Rogers, J., Huang, Y., 2008. Quantification of employment from biomass power plants. Renew. Energy 33 (8), 1922–1927.

Toke, D., 2007. Renewable financial support systems and cost-effectiveness. J. Clean. Prod. 15 (3), 280–287.

Trink, T., Schmid, C., Schinko, T., Steininger, K.W., Loibnegger, T., Kettner, C., Pack, A., Töglhofer, C., 2010. Regional economic impacts of biomass based energy service use: a comparison across crops and technologies for East Styria, Austria. Energy Policy 38 (10), 5912–5926.

Van Dam, J., Faaij, A.P., Hilbert, J., Petruzzi, H., Turkenburg, W., 2009. Large-scale bioenergy production from soybeans and switchgrass in Argentina: part B. Environmental and socio-economic impacts on a regional level. Renew. Sustain. Energy Rev. 13 (8), 1679–1709.

Van den Broek, R., Van den Burg, T., Van Wijk, A., Turkenburg, W., 2000. Electricity generation from eucalyptus and bagasse by sugar mills in Nicaragua: a comparison with fuel oil electricity generation on the basis of costs, macro-economic impacts and environmental emissions. Biomass Bioenergy 19 (5), 311–335.

Van der Horst, D., Vermeylen, S., 2011. Spatial scale and social impacts of biofuel production. Biomass Bioenergy 35 (6), 2435–2443.

Van Eijck, J., Romijn, H., Smeets, E., Bailis, R., Rooijakkers, M., Hooijkaas, N., Verweij, P., Faaij, A., 2014a. Comparative analysis of key socio-economic and environmental impacts of smallholder and plantation based jatropha biofuel production systems in Tanzania. Biomass Bioenergy 61, 25–45.

Van Eijck, J., Romijn, H., Balkema, A., Faaij, A., 2014b. Global experience with jatropha cultivation for bioenergy: an assessment of socio-economic and environmental aspects. Renew. Sustain. Energy Rev. 32, 869–889.

Walsh, M.E., Daniel, G., Shapouri, H., Slinsky, S.P., 2003. Bioenergy crop production in the United States: potential quantities, land use changes, and economic impacts on the agricultural sector. Environ. Res. Econ. 24 (4), 313–333.

Weldegiorgis, F.S., Franks, D.M., 2014. Social dimensions of energy supply alternatives in steelmaking: comparison of biomass and coal production scenarios in Australia. J. Clean. Prod. 84, 281–288.

White, W., Kulisic, B., Domac, J., 2007. Economic and social drivers to encourage bioenergy market development'. Proceedings of Fifteenth European Biomass Conference and Exhibition, 7–11 May, 2007, Berlin, Germany.

White, W., Patriquin, M., 2003. A regional economic impact modelling framework. Paper submitted to the XII World Forestry Congress, 2003, Quebec City, Canada, Available from: http://cfs.nrcan.gc.ca/pubwarehouse/pdfs/26898.pdf

Wicke, B., Faaij, A., Smeets, E., Tabeau, A., Hilbert, J., 2007. The socio-economic impacts of large-scale land use change and export-oriented bio-energy production in Argentina. Proceedings of the Fifteenth European Biomass Conference and Exhibition, 7–11 May, 2007, Berlin, Germany.

Winfield, M., Dolter, B., 2014. Energy, economic and environmental discourses and their policy impact: the case of Ontario's Green Energy and Green Economy Act. Energy Policy 68, 423–435.

Chapter 7

Challenges and Opportunities for the Conversion Technologies Used to Make Forest Bioenergy

William Cadham, J. Susan Van Dyk, J.S. Linoj Kumar, Jack N. Saddler
Forest Products Biotechnology/Bioenergy Group, University of British Columbia Vancouver, BC, Canada

Highlights

- The majority of globally harvested wood is used for energy generation, with 60% employed for traditional (open fire/charcoal production), inefficient combustion.
- Modern bioenergy is also dominated by combustion, during which forest-derived biomass is converted in highly efficient boilers, furnaces and stoves.
- Technologies such as pyrolysis, gasification and biochemical conversion can create more energy-dense and adaptable products, enhancing forest biomass mobilisation. These technologies will be important in the development of biomass-based transportation fuels.
- Modern biomass conversion technologies offer enormous opportunity, in both developing and developed countries, through better use of forest and mill residues, improved efficiency of biomass conversion pathways, reduced energy-related carbon emissions and enhanced energy security.

INTRODUCTION

It is sometimes wrongly assumed that the majority of the world's annual forest harvest is used to produce wood products such as lumber, panel, furniture, pulp, paper and engineered wood products. On a global level, the biggest share (55%) of forest fibre is actually used to produce energy, rather than wood products. Demand for traditional fuelwood volumes alone are greater than the demand for industrial roundwood, as illustrated in Fig. 7.1 (Chum et al., 2011). In the developing world, biomass is mainly used for fuel in traditional applications, such as cooking, charcoal production and heating. Although developed countries generally convert most of their annual forest harvest into wood products,

Mobilisation of Forest Bioenergy in the Boreal and Temperate Biomes

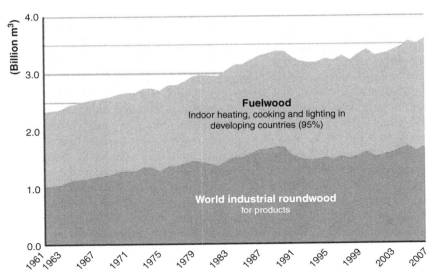

FIGURE 7.1   Fuelwood used in developing countries parallels world industrial roundwood production levels. *(Source: From Chum et al. (2011), Fig. 2.1, bottom.)*

there are also several countries where forest biomass plays an important role in energy generation, such as Finland and Sweden (also see chapter: Comparison of Forest Biomass Supply Chains From the Boreal and Temperate Biomes). In this chapter, we review the use of forest biomass for bioenergy production, discuss the technologies used for conversion of forest biomass and examine some of the key jurisdictions where forest biomass is utilised for modern bioenergy applications.

The most common method for converting forest biomass into energy is direct combustion, which can be simply defined as the burning of biomass. In the developing world, wood is typically burned in open fires or simple stoves for low-grade heating and cooking. These more traditional bioenergy applications achieve very low conversion efficiencies (about 10–20% of the intrinsic energy of biomass), but account for 60% of bioenergy production across the globe, and 35% of final energy demand in developing countries (Chum et al., 2011; IEA, 2012a). Some regions, such as sub-Saharan Africa, are heavily reliant on bioenergy, which accounts for 61% of primary energy demand (IEA, 2013a).

Modern bioenergy applications are also dominated by combustion, but use much more efficient technologies to generate heat and combined heat and power (CHP). Although these processes only supply about 3% of the developed world's energy demand (Dornburg et al., 2010), this is projected to increase in the future. The International Energy Association's (IEA) Technology Roadmap (IEA, 2012a) suggests that, by 2050, biomass could supply 160 EJ of energy, of which 100 EJ will be from bioenergy and 60 EJ from biofuels. In 2010, the contribution of biomass to the world's energy demand was ~53 EJ. Thus, to achieve the IEA's projected targets, the world will require access to 8–11 × 10⁹ dry tonnes of biomass per year, generated from a range of sources, including mill wastes, forest residues and energy crops.

Compared to fossil fuel feedstocks, forest biomass has a low energy density and is primarily consumed locally. In contrast, large volumes of fossil fuels are traded globally: 900 million tonnes of coal was traded in 2011 (26.4 EJ) (IEA, 2013a). Although wood pellets are also traded globally, the international pellet market represents only a small fraction of the world's total demand for biomass-based energy. The quantities of forest biomass harvested for energy production in Finland and Sweden are some of the highest in the world (Hakkila, 2006; IEA, 2013b), yet the demand for bioenergy has expanded to such an extent that even these countries are importing pellets (IEA, 2013b). Other countries, such as Canada and the United States, have an even larger supply of forest fibre, with an average of 170 and 422 million cubic metres (Mm³) of roundwood harvested between 2000 and 2012, respectively (see chapter: Comparison of Forest Biomass Supply Chains From the Boreal and Temperate Biomes). In contrast to Finland and Sweden, however, bioenergy applications are limited in Canada and the United States, in large part because these countries are relatively well-endowed with less expensive fossil fuels and hydro-electricity (FAO, 2013). The availability of low-cost alternatives makes it difficult for bioenergy to become cost-competitive without government policies, such as carbon pricing, that discourage fossil fuel consumption. Unlocking biomass potential in regions such as Canada and the United States, which have an abundant supply of woody biomass that is not fully utilised, presents both challenges and opportunities for bioenergy conversion technologies.

With respect to evolving technologies, such as pyrolysis, gasification and biochemical conversion, it is likely that combustion will remain the most prevalent bioenergy process for the immediate future. However, traditional/developing world bioenergy applications are expected to decline, as modern technologies are further developed and as poorer countries' standards of living increase, such that they can better access fossil fuel or alternative renewable energy technologies (IEA, 2013a). Agricultural biomass can be used for some bioenergy applications, for example for wheat straw-based CHP facilities or biogas production. However, there remain considerable challenges with the use of agricultural-based feedstocks (such as seasonal availability, low density and, typically, a higher ash content), which make them less desirable as bioenergy feedstocks. If forest biomass is to become more universally traded, it will likely require the production of higher-density intermediate products, such as pellets and bio-oils, to facilitate the movement of bioenergy products from regions with abundant forest biomass to those with limited access to forest biomass.

FOREST BIOMASS FEEDSTOCKS AND CONVERSION TECHNOLOGIES

Forest biomass can include forest and mill residues as well as specially grown woody crops, such as poplar and willow. Poplar and willow can be very effective energy crops because they grow quickly in dense plantations (10,000–20,000 plants/ha), can be harvested in short rotations (3–5 years) and remain productive for seven or eight harvests. This type of growth system is known as short-rotation coppice (SRC) and is prevalent across Europe and in parts of North America. Major SRC regions include Sweden (~14,000 ha of willow), Italy (~6000 ha, dominated by poplar), Poland (~3000 ha, dominated by willow), the United Kingdom (~7 500 ha, dominated by willow) and Germany (~5000 ha of poplar and willow) (Langeveld et al., 2012). In North America, SRC plantations of willow and poplar are concentrated in the east, primarily in New York State and in the Canadian provinces of Quebec and Ontario. Unlike European plantations, eastern North American plantations are employed principally for research purposes. Poplar plantations in western North America can also be found in southern British Columbia, Washington and Oregon. In Oregon, ~14,600 ha of hybrid poplar plantations have been planted, with an average production rate of 12.9 million tonnes ha/year (dry) (Berguson et al., 2010). By comparison, Brazilian sugar plantations covered ~9.8 million ha in 2012, with an average production rate of ~67 Mtonnes/ha per year (UNICA, 2014).

As mentioned in the chapter: Introduction, forest biomass can include primary and secondary residues. Primary residues include those from conventional forest management, such as thinnings, slash and stumps, while secondary residues are by-products from industrial processing, such as sawdust, wood chips and black liquor (Röser et al., 2008). Global estimates of potential forest biomass residues range between 10.0 and 15.9 EJ/year, although projections can vary by an order of magnitude (Haberl et al., 2010; Hoogwijk et al., 2003) (also see chapter: Quantifying Forest Biomass Mobilisation Potential in the Boreal and Temperate Biomes). Both specialised energy crops and forest residues have a low energy density compared with conventional fossil fuels and this factor has limited the expansion of biomass for energy generation to date.

Biomass Upgrading

The low energy density of forest biomass results in high handling/transport costs, low conversion efficiencies and, ultimately, an inflated price for feedstocks (Mabee and Mirck, 2011). To address this problem, various biomass-upgrading methods, such as pelletisation, torrefaction, pyrolysis, gasification and biochemical conversion, have been developed.

Pelletisation

Pelletisation is a commercial technology that involves the mechanical compression of bulky biomass, primarily sawdust, to create wood pellets with an energy density of 11 GJ/m^3 (low heating value, LHV) (IEA, 2012a; Stelte et al., 2012; Tumuluru et al., 2011). Wood pellets are internationally traded as a bioenergy feedstock (Box 7.1).

BOX 7.1 Wood Pellets and International Mobilisation

Wood pellets are the fastest growing of any bioenergy feedstock. In 2010, global pellet production reached 14.3 million tonnes, while consumption reached 13.5 million tonnes. Compared to global production in 2006 (~6–7 million tonnes), this represents a 110% increase (Goh et al., 2013). In general, pellets can be classified as either: high-quality pellets (white pellets, supplied in bulk or in bags) destined for the residential heating market, or industrial-quality pellets (brown pellets, supplied in bulk) made from lower-value feedstocks and destined for district heating systems (DHS) or power and CHP operations (Goh et al., 2013; Mabee and Mirck, 2011).

Global trade of wood pellets (45 PJ) is comparable to that of biodiesel (75 PJ) and bioethanol (16–22 PJ) (Goh et al., 2013). The European Union is the main market for both white and brown wood pellets; consumption of wood pellets in the EU reached 11.4 million tonnes in 2010, and accounted for 85% of global demand. European pellet production satisfied ~80% of continental demand in 2010, but was heavily reliant on trade within the EU: 4.1 million tonnes of wood pellets were traded among EU states, from regions with high pellet production relative to demand, such as the Baltic states, to regions such as the United Kingdom. That same year, the EU imported 2.6 million tonnes of wood pellets, primarily from Canada (0.9 million tonnes) and the United States (0.4 million tonnes) (Goh et al., 2013; Sikkema et al., 2011). The production of wood pellets in Western Canada for export to the European market is currently being challenged by expanding production capacity in the US Southeast, where shipping and handling costs are lower. However, increased demand for wood pellets in Japan and South Korea may create a new market for Western Canadian pellets in the next few years (Goh et al., 2013). See chapter: Challenges and Opportunities for International Trade in Forest Biomass for more details on trade.

More recently, torrefaction has been explored as a way of making pellets more easily integrated with coal distribution and use. Torrefaction is similar to traditional charcoal production and involves the heating of biomass under anaerobic conditions to between 200 and 300°C (Acharya et al., 2012; Chew and Doshi, 2011; Ciolkosz and Wallace, 2011). Torrefaction increases the bulk density of wood pellets by 25–30%, compared to conventional wood pellets, and limits the risk of self-heating and spontaneous combustion, which can sometimes occur when wood pellets undergo biological degradation. Torrefaction can also, if processed appropriately, improve pellet durability during transport and allow for outdoor storage with very limited risk of degradation compared to non-torrefied pellets (Li et al., 2012). Torrefaction is currently at a pre-commercial stage, and requires further process optimisation before it will be fully commercialised (Batidzirai et al., 2013).

Pyrolysis

During pyrolysis, organic material is heated to temperatures greater than 400°C in the absence of oxygen to create bio-oil (also known as pyrolysis oil), biochar and synthetic gas. The percentage yield of each product depends on the specific reaction conditions used, with fast pyrolysis creating higher levels of bio-oil. Various reviews on this technology have been published, and technological improvements are an ongoing subject of research (Bridgwater, 2012; Butler et al., 2011; Mohan et al., 2006).

Pyrolysis oil (19 GJ/m^3 LHV) has an energy density nearly twice that of conventional wood pellets and is, therefore, more practical for long distance transportation (IEA, 2012a). However, the oil is unstable and deteriorates quickly when stored. Pyrolysis oil also has the potential to serve as a feedstock for chemical and transportation fuel applications, but commercialisation of advanced applications is currently limited by the requirement that the oil undergo extensive processing and upgrading with hydro-treatment. Consequently, significant cost reduction and optimisation are required before these materials can be used widely (Arbogast et al., 2013).

Gasification

Gasification typically involves the thermochemical conversion of carbon-based materials to produce synthetic gas, or syngas. The technology has been used extensively for fossil fuel feedstocks, where gasification is carried out at high temperatures (>900°C) in the presence of air, steam and oxygen. Syngas is a mixture of carbon monoxide and hydrogen, as well as smaller quantities of methane, carbon dioxide and nitrogen. Syngas can serve as a flexible intermediate that can be upgraded through catalysis to multiple chemical products and fuels through the Fischer–Tropsch (FT) process. Gasification technologies can be applied in power and heat generation, as well as in the production of transportation fuels and gases. For example bio-methane has similar properties to natural gas and can be integrated into the natural gas grid, gas power plants and the transportation sector (Damartzis and Zabaniotou, 2011; IEA, 2012a). Various reviews have been published on the gasification of biomass and its technical challenges (Ahrenfeldt et al., 2013; Balat et al., 2009; Göransson et al., 2011; Kirkels and Verbong, 2011; Pereira et al., 2012). Gasification using municipal solid waste (MSW) as a low-cost feedstock has become more prevalent, with several commercial facilities currently under construction (Arena, 2012). Although FT synthesis of advanced chemicals and fuels are fully commercialised for feedstocks such as coal and natural gas, this is not the case for biomass-derived gasification. Biomass feedstocks contain high levels of oxygen and impurities that lead to tar formation and require different processing and catalysts than fossil fuel feedstocks. Thus, extensive optimisation and improvement is required before this processed will be commercialised for bioenergy or biofuel applications (Karatzos et al., 2014).

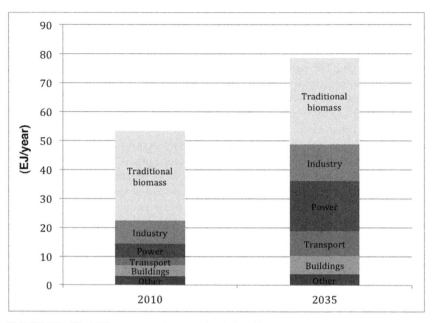

FIGURE 7.2   World bioenergy use by sector in 2010 and 2035 (IEA, 2012b).

Biochemical Conversion

Biochemical conversion of biomass into liquid fuels typically involves the use of enzymes or microbial catalysts to hydrolyse carbohydrates into simple sugars. Although liquid fuel (ethanol) production from starch and sugars is a fully commercialised process, cellulosic ethanol is just entering the commercial stage (Chundawat et al., 2011; Jørgensen et al., 2007; Menon and Rao, 2012). Currently, cellulosic ethanol plants use agricultural residues as their major feedstock, primarily because they are located close to existing ethanol facilities that use corn, wheat or sugar cane. However, it is likely that hardwoods, due to their inherent characteristics, could also be readily adapted as suitable feedstocks for cellulosic ethanol production.

BIOMASS APPLICATIONS AND CONVERSION TECHNOLOGIES

Biomass can be converted into a number of energy products and is, therefore, widely employed throughout the energy sector. The use of biomass in 2010 and projections for use in 2035 are illustrated in Fig. 7.2. Currently, most biomass is used for traditional bioenergy purposes (60%), followed by industry (15%) and power (9%), while transport, buildings and other uses of biomass account for 5, 7 and 6%, respectively. By 2035, the use of biomass within the energy sector is projected to change significantly. The most obvious change will be a decline in traditional (developing world) bioenergy applications, which are expected to fall from ~60 to 37%, by 2035, as access to modern conversion technologies improves (IEA, 2012b). The IEA predicts that advanced and drop-in biofuels (ie fuels that can be used without major investments in infrastructure) will only provide significant contributions to the transportation sector after 2035 but will increase significantly through 2050 (IEA, 2012b).

Changing energy requirements will create new demands on all aspects of the bioenergy supply chain, including harvest practices, transportation and conversion technologies. The remainder of this section examines the utilisation of biomass for different applications within the energy sector, including the use of biomass for energy in industrial applications, power production, buildings and the transportation sector. The primary conversion technologies used to convert forest biomass—and the role that these technologies will play within the energy sector—are also discussed.

Forest Biomass in the Energy Sector

In the developed world, facilities integrated within industrial processes such as pulp and paper production currently generate most of the energy from forest biomass (Sixta, 2008). Combustion of biomass for the production of industrial heat and CHP have been used by the forest harvesting and wood manufacturing industries since the 1930s, when the Tomlinson recovery boiler was developed (Sixta, 2008). Although industrial demand for bioenergy is projected to increase by only 1% (from 15 to 16% of total bioenergy demand) by 2035, this accounts for ~4.6 EJ, or more than one-third of the installed capacity in 2010 (Fig. 7.2). Nevertheless, biomass use for industrial energy production will probably have only a marginal impact on global biomass mobilisation in the future.

Outside of industry, stand-alone facilities exist where forest biomass is consumed to produce heat, power and CHP. In Fig. 7.2, biomass for heat and power are denoted as 'buildings' and 'power', respectively. Of all biomass-based energy uses, power production is expected to increase the most: from ~4.8 EJ in 2010 to ~17.3 EJ in 2035 (IEA, 2012b). Over this same period, the use of biomass for heating in buildings is expected to increase from ~3.7 to ~6.3 EJ. These projections are driven by expectations of increased combustion of biomass in CHP facilities and co-firing with coal to help countries meet renewable energy and GHG emissions targets (IEA, 2013a).

The massive expansion in the use of biomass for power and heat in recent years (and the expectation that this trend will continue) has been a driving force behind the increased use of forest bioenergy. For example, growth in the residential pellet market in Scandinavia (discussed later in the chapter), Germany and Austria has resulted in the reallocation of biomass residues for white pellet production. At the same time, increased demand for industrial pellets has been driven by the construction of large-scale biomass power facilities, such as Drax (4000 MW_{el} requiring 7.5 million tonnes of biomass annually). It is important to note, however, that power generation from biomass can be over-estimated because the entire electrical output from co-firing facilities is often attributed to bioenergy, even though some fraction of the energy produced is actually derived from coal.

The use of biomass within the transportation sector is expected to increase more than threefold, from ~2.6 EJ in 2011, to ~8.6 EJ in 2035 (IEA, 2013a). As mentioned earlier, sugar- and starch-rich biomass feedstocks dominate the production of liquid transportation fuels via biochemical conversion. With the production of transportation biofuels expected to increase, there will be an impact on biomass mobilisation in the future.

Bioenergy Conversion Technologies for Forest Biomass

Combustion of forest biomass for traditional bioenergy production is, and will remain, the most prevalent conversion technology through to 2035, as illustrated in Fig. 7.2 (IEA, 2012b). Modern applications of bioenergy are also dominated by combustion, but highly efficient technologies are used to generate heat, power and CHP. These modern technologies are the focus of the following section.

Biomass for Heat Generation

Domestic Heating and Cooking

The most widespread use of bioenergy, concentrated in developing countries in Africa and Southeast Asia, is traditional combustion for domestic heat and cooking. Open fires or crude stoves with conversion efficiencies of only 10–20% are typically used. Major problems with this type of bioenergy generation include unsustainable sourcing of biomass, severe smoke pollution and emissions with considerable global warming potential (Bauen et al., 2009).

Combustion of biomass in residential buildings for heat is also common in developed countries, although the technologies applied and the feedstocks employed can vary significantly. Domestic heat applications ($\sim$1–5 MW_{th}) include open fires and wood stoves that burn fuelwood (10–20% conversion), or modern furnaces, boilers and stoves that burn fuelwood or pellets (70–90% conversion) (Faaij, 2006). Traditional, open-flame combustion is still prevalent in some developed countries, despite the host of modern technologies available to improve the efficiency of combustion. In 2011, France was the largest consumer of fuelwood of all EU member states (25.8 Mm^3), followed by Denmark (11.1 Mm^3). This fuelwood was consumed primarily for domestic heating applications in rural towns (FAO, 2013). By contrast, modern combustion technologies are widely deployed in households across Scandinavia, Austria and Germany (IEA, 2013b; Lipp, 2007; Schmidt et al., 2011).

District Heating and Cooling

DHS involves the combustion of biomass feedstocks in a single facility to produce steam, which is subsequently forced through a network of pipes and used to heat buildings (Bauen et al., 2009). Supplying a significant amount of the heating services in some, predominantly northern, regions ($\sim$50% in Denmark and Finland), large-scale combustion (5–10 MW_{th}) for heat is a mature technology. Generated heat is cost-competitive with that produced from fossil fuels. However, DHS require a significant investment in infrastructure. Furthermore, operations are restricted to the cooler months, when heating is required, hindering the deployment of new facilities, as installations and equipment would remain unused for part of the year. On the other hand, it should be noted that interest in absorption cooling technologies is on the rise, because they are more efficient and can utilise DHS infrastructure year-round (Bauen et al., 2009).

Industrial Heating

It is common for facilities such as pulpmills, with a high heat demand and large quantities of available biomass, to employ boilers (0.5–10 MW_{th}) that can combust forest and mill residues. Because production of heat alone typically results in less effective energy recovery, compared with co-generation of heat and power, many industrial heating systems are transitioning to CHP generation.

Biomass for Power and CHP Generation

Although multiple feedstocks and conversion technologies are available to convert biomass into power or CHP, the predominant technologies used by biomass-based power plants, co-firing facilities and biomass-based cogeneration plants are briefly reviewed.

Biomass-Based Power Plants

The most established method of power production from biomass in stand-alone facilities is the steam-powered turbine (IEA, 2012a). In this system, combustion of biomass in a boiler produces heat, which is subsequently employed to generate steam that powers a turbine to produce electricity. The efficiency of the process is heavily contingent upon the facility's scale, while the scale is typically reliant on the availability of low-cost feedstocks. Because biomass prices rise sharply with increasing distance between facility and suppliers, a balance is required to maximise the scale of a facility, while minimising feedstock transport distances. Typically, large-scale 30–100 MW_{el} facilities are required to make biomass-based steam-powered turbine processes economically viable. Box 7.2 provides a brief case study of the single largest biomass power facility in North America. Plants of this size can achieve efficiencies in the 18–33% range, somewhat lower than fossil fuel facilities of a similar scale (IEA, 2012a). It should be noted that the viability of a number of small (5–10 MW_{el}) steam facilities is increasing across Europe and North America, as new feedstocks, primarily forest residues, are becoming available at relatively low cost (Vakkilainen et al., 2013).

BOX 7.2 Williams Lake Bioenergy Facility, British Columbia, Canada

In 1993, primarily motivated by severe smoke pollution caused by inefficient burning of mill and logging residues, EPCOR built the Williams Lake Generating Station. It remains the largest continually operating, dedicated biomass power facility in North America. Located in Williams Lake, BC, the 66 MW_{el} facility uses 600,000 tonnes of wood residues annually to produce electricity (Nicholls et al., 2008). Bark, chips and sawdust residues are supplied by five local sawmills, all located within ~5 km of the power facility. The feedstock is provided at no cost, although EPCOR pays for truck transport of the residues and invested approximately 2 million CA$ for residue preparation equipment at each mill (Nicholls et al., 2008). The conversion efficiency would be enhanced if the steam produced from the facility was also used to supply industrial processes, or was fed into a DHS grid, using CHP.

Co-Firing

The combined combustion of biomass with fossil fuels to create heat and power (ie CHP) is one of the most cost-effective methods of transforming biomass into electricity and heat, when the required infrastructure exists. Biomass co-firing with coal has emerged as the most popular method of CHP, with over 150 facilities worldwide (Al-Mansour and Zuwala, 2010). Co-firing leverages the existing infrastructure of coal power plants, offering the opportunity to increase the proportion of renewables in the primary energy mix and reducing the GHG emissions associated with coal combustion. This has successfully encouraged the use of co-firing facilities across Europe, most notably in Scandinavia, Germany, Belgium, the Netherlands and the United Kingdom. The United Kingdom is the home of Drax, the world's largest biomass using power facility (4000 MW_{el}).

Three basic systems are used to co-fire biomass with coal. These are direct, indirect and parallel co-firing, all of which have been proven at an industrial scale (Vakkilainen et al., 2013). In direct co-firing, biomass is combusted with coal in proportions ranging from 5–10%, simply by mixing the two fuels. Direct co-firing requires only a minor capital investment for biomass pretreatment and feed-in systems. However, relatively homogenous feedstocks are required. Forest biomass-based industrial pellets are an ideal feedstock for direct co-firing, because they provide the homogeneity required for optimal facility performance. By contrast, the lack of homogeneity and high ash content of agricultural residues create challenges with direct co-firing operations, effectively limiting their suitability as a feedstock. Although indirect or parallel co-firing techniques (in which biomass and coal are fed into the boiler separately) can be employed to abate the challenges of agriculture feedstocks, these systems are more capital-intensive than direct co-firing (Fernando, 2009). Indirect co-firing often involves the gasification of solid biomass prior to its combustion with coal, offering high feedstock flexibility and the ability to clean the fuel gas prior to combustion, thereby minimising damage to the boiler and improving the longevity of the infrastructure. This type of system is currently employed in the 167 MW$_{el}$ Lahti plant in Finland, where 17% biomass is combusted with a mixture of coal and natural gas (Al-Mansour and Zuwala, 2010).

Biomass-Based Co-Generation

Co-generation can significantly increase the overall efficiency and competitiveness of a stand-alone biomass-based power plant or industrial facility (eg a pulp mill) by capturing the 'waste' heat. Typical overall (thermal + electrical) efficiencies are within the range of 80–90% (IEA, 2012a). In stand-alone CHP facilities, the low-grade steam left over after power production is used to provide heat for residential or commercial buildings, where DHS infrastructure is available. Heat recycling is an important step in biomass co-generation, and has been shown to reduce power production costs by 40–60% in stand-alone facilities within the 1–30 MW$_{el}$ range, by recovering the exhaust heat from electricity generators and employing it to create steam (IEA, 2012a). In domestic and commercial applications, the scale of CHP facilities is constrained by the heat demand within the region it is servicing. This is further complicated by the seasonal nature of this demand (see discussion about seasonality in chapter: Challenges and Opportunities of Logistics and Economics of Forest Biomass). As is the case with district heating, the growing interest in supplementing CHP systems with cooling technologies may further improve the efficiency and economics of the process.

Industrial biomass-based CHP facilities typically combust residues such as black liquor (a by-product of the Kraft pulping process) in order to generate electricity and low-grade steam (Sixta, 2008). The power and excess steam produced is employed upstream within the processing facility. Given that the Tomlinson recovery boiler has been around since the 1930s, this process has been part of best practices in Kraft pulpmills for a number of decades. Combustion of black liquor originally emerged as a way to recycle expensive pulping chemicals, while the benefits of heat and power production were only realised more recently (Sixta, 2008). A number of government programmes have provided subsidies and grants to improve the recovery systems within chemical pulpmills. As described in Box 7.3, various policies have been used to support black liquor combustion in Canada and the United States. Indeed, the largest share of bioenergy production in North America comes from the operation of recovery boilers at Kraft mills. In the United States, 60% of all wood used for bioenergy occurs within industry. In Canada, the proportion is even higher: approximately 80% of all Canadian biomass-based energy ($\sim$400 PJ of the total 516 PJ) is derived from the industrial processing of wood waste and spent pulp liquor (Bradley and Bradburn, 2012).

BOX 7.3 US Black Liquor and Canadian Pulp and Paper Green Transformation Programmes

After the passage of the US Highway Act of 2005, owners of US Kraft pulp-mills began exploiting a tax credit 'loop hole' originally designed to promote renewable fuels in the transport sector. Prior to combustion in the recovery boiler, addition of 0.1% diesel fuel to the black liquor was sufficient to deem the mixture a 'n' and qualify for a 0.50 US$/gallon tax credit. Pulpmills that took part in the programme received financial pay-outs of 6 billion US$ in 2009 alone, effectively subsidising the US pulp industry. Estimates suggest that the black liquor subsidy translated into a chemical pulp price reduction of ∼ 300 US$/tonne. The policy was seen as a failure because it awarded the pulp and paper industry for increasing their consumption of diesel, which was not traditionally blended with black liquor and effectively raising GHG emissions from pulpmills.

Canadian pulpmills experienced an immediate negative financial impact because they were unable to compete with the artificially low pulp prices of their US counterparts. The Canadian federal government struggled to respond because any direct government subsidy violates World Trade Organization standards. In 2009, the Government of Canada created the 1 billion CA$ Pulp and Paper Green Transformation Program (PPGTP). Under PPGTP, credits valued at 0.16 CA$/L were provided for black liquor produced at Canadian pulpmills, to a maximum pay-out of CA$1 billion; generated credits could be used to finance capital projects aimed at reducing the environmental impact of the pulp sector. Eligible projects included investments in energy efficiency, boiler upgrades and turbine instillations. The fund was fully utilised within 4 months. The government claims the programme:

- generated enough renewable electricity to power 140,000 homes;
- produced enough renewable thermal energy to heat 70,000 homes;
- saved enough energy to heat 135,000 homes;
- cut GHG emissions from the Canadian pulp and paper sector by 12% of 2009 levels; and
- reduced water use, waste and air pollution.

MAJOR BIOMASS-TO-ENERGY USERS

As mentioned in the Introduction to this chapter, biomass is primarily consumed locally. As a result, the importance of bioenergy varies widely between countries with an abundance of available biomass and those with a shortage. In this section, four countries in the temperate and boreal biomes (namely Finland, Sweden, Canada and the United States) that play a major role in the forest biomass-to-energy industry are discussed, in the context of their biomass availability, renewable energy and GHG mitigation targets, as well as the availability and cost of other sources of energy.

Finland

Although forest-derived biomass has always been an important source of energy in Finland, it was only as oil prices skyrocketed in the 1970s that biomass re-emerged as one of the county's primary energy sources. In 2013, forest biomass accounted for a larger proportion (24%) of final energy consumption than any other fuel, including oil (Statistics-Finland, 2014).

Finland is endowed with considerable forest resources. Land ownership is characterised by a high private share (61%), limited corporate holdings (9%), some state ownership (25%) and the remainder (5%) owned by public interest groups. The majority of private forest owners belong to management associations that provide professional assistance with forest operations. These cooperative organisations have a significant impact on the development of bioenergy in the country because organised private owners can both influence and respond to policies shaping the forest industry (Ericsson et al., 2004).

Approximately 42 million m^3 of solid wood was consumed for energy production in Finland in 2013; this includes firewood, bark, sawdust, pellets, wood chips, demolition wood, briquettes and pellets (Statistics-Finland, 2014). Finland has the third highest capacity of biomass power generation in the EU, after Germany and Sweden. Co-generation facilities in the 2–20 MW_{el} range provide most of Finland's heat and power, including more than 30% of all electricity generation (Salomón et al., 2011). Finnish CHP facilities are a mixture of stand-alone operations and those integrated within industry. Surplus heat produced from CHP plants are fed into the nation's widespread DHS grids. District heating accounts for ~50% of total heat demand, ~75% of which originates in CHP facilities.

Although some of the heat and power generated in CHP facilities is consumed in residential and commercial applications, the major consumer of bioenergy is industry. The forest products sector has operational infrastructure that bioenergy can leverage, lowering the barriers to production. In Finland, the wood products and manufacturing sector is the single largest producer and consumer of bioenergy: in 2010, this sector accounted for ~70% of all bioenergy production, the lion's share of which was employed in upstream processes (Aslani et al., 2013). Combustion of black liquor for chemical recovery in Kraft pulpmills is the most important source of bioenergy in Finland (IEA, 2013b).

Sweden

In Sweden, the timeline for bioenergy development resembles that of Finland, where forest biomass was historically an important fuel source until it was surpassed by oil. The absence of significant domestic fossil fuel resources in Sweden meant that the country became reliant on imports for most of its energy supply. A reduction in the country's oil dependence became a high priority after the 1973 oil crisis, and biomass emerged as a viable alternative. Today, bioenergy is the largest component of the Swedish national energy mix: in 2013, biomass accounted for ~34% of the country's total energy consumption, the highest proportion of any OECD country (SVEBIO, 2013). Forest biomass supplies ~90% of the bioenergy feedstock in Sweden.

In Sweden, as in Finland, the forest ownership structure and presence of a strong forest industry has promoted the rapid penetration of bioenergy into the national energy mix. Forest ownership includes a high proportion of private (49%) and corporate (40%) owners, improving the ability of the forest industry to influence and respond to government policy. The forest ownership structure in Sweden (and Finland) is an important reason for the success of the forest bioenergy industry and differentiates them from the other countries (Ericsson et al., 2004). In 2013, Sweden remained the world leader in bioenergy generation: biomass accounted for 33% of the country's final energy mix (0.46 EJ) (SVEBIO, 2013). Forest biomass, including black liquor, is the predominant bioenergy feedstock in Sweden (Salomón et al., 2011).

DHS have been enormously influential in the advancement of bioenergy in Sweden, providing 93 and 83% of heating services in apartment buildings and commercial spaces, respectively (SEA, 2013). Unlike Finland, the potential for Sweden to produce electricity from CHP has been underutilised, primarily because of competition with low-cost alternatives, such as nuclear and hydropower. In 2003, the Swedish government introduced Green Electricity Certificates (GEC) to support the production of renewable energy (Westholm and Lindahl, 2012). The implementation of GEC has helped to spur a transition from district heating to CHP. By the end of 2015, installed CHP capacity in Sweden is expected to reach 1250 MW_{el}, or ~15% of electricity generation (Salomón et al., 2011).

Despite the prominence of forest bioenergy in residential and commercial applications, industry is still the largest consumer of forest biomass for energy in Sweden. In 2010, industrial energy demand (~0.53 EJ) accounted for 24% of total Swedish energy use, 42% (~0.22 EJ) of which originated from biomass, including black liquor (0.14 EJ), other pulp residues (~0.04 EJ) and sawmill residues (~0.01 EJ) (SEA, 2013). The use of bioenergy for industrial applications represents 50% of total Swedish bioenergy demand. The pulp and paper industry is the major consumer of energy; in 2010, this sector accounted for 52% (~0.28 EJ) of total industrial energy demand. The pulp and paper industry also accounts for approximately 90% of industrial bioenergy production (~0.18 EJ, excluding biomass-based electricity from the grid) (SEA, 2013).

Sweden has also been one of the most successful European countries in promoting the use of renewable transportation fuels: in 2013, biofuels accounted for 9.7% of total transportation fuel demand (Grahn and Hansson, 2015). The market is dominated by bioethanol, biodiesel and biogas. Although agricultural residues are the primary feedstock for biofuels currently consumed within Sweden, significant investment in research and development of technologies to transform forest biomass into transportation fuels is ongoing. For example, the production of bio-oil is being explored at the SunPine facility in Piteå (Grahn and Hansson, 2015). This product is subsequently converted into biodiesel at a refinery in Gothenburg. Forest-derived transportation fuels have been available on the Swedish market since 2011, and have an annual capacity of 1 million litres (Grahn and Hansson, 2015).

The United States

Biomass is the largest source of renewable energy in the United States: in 2011, biomass accounted for 48% (4.6 EJ) of total renewable energy production. Forest-derived biomass is the most commonly used feedstock of all biomass sources, despite the large volumes of biofuels produced from corn (EIA, 2012a).

According to the US Energy Information Agency (EIA), wood and biomass waste in the United States is used for heat and power production, industrial, residential and commercial operations and electricity production in stand-alone facilities (EIA, 2012a). The industrial sector is the largest consumer of biomass for bioenergy production in the United States: in 2011, the industrial sector consumed ~60% of all the wood and waste biomass used for bioenergy. Industrial bioenergy consumption is concentrated within the recovery boilers of Kraft pulp-mills. In 2009, the industrial sector in the United States consumed 2.09 EJ of the ~2.5 EJ consumed by all sectors together (EIA, 2009). Pulp, paper and associated industries were the largest consumers of biomass for energy, using 49.7% (1.04 EJ) of industrial bioenergy; the largest share of this energy (0.72 EJ) was derived from the combustion of black liquor. Eighty five per cent of the bioenergy consumed in the pulp sector is used for heat production (0.88 EJ), with the remainder (0.16 EJ) for electricity. Of the electricity generated by the pulp sector, ~54% (0.087 EJ) was sold to the grid. Generated heat is recycled upstream in the pulping process, thereby reducing the energy requirements of the mill.

Biorefineries are the second largest consumers of biomass in the United States, accounting for 31.1% (0.65 EJ) of industrial bioenergy, while the lumber manufacturing industry combusts wood waste to generate 10.5% (0.22 EJ) of industrial bioenergy (EIA, 2009, 2012b). The use of stand-alone DHS and CHP facilities in the United States is limited because much of the country lacks centralised heating infrastructure. Use of woody biomass for energy and heat is, thus, restricted primarily to residential and small commercial heating applications and to electricity-only operations. The use of biomass for residential and commercial space heating in the United States generally involves combustion of fuelwood or wood pellets in traditional open fires, or higher efficiency stoves (White, 2010). Together, residential and commercial bioenergy applications accounted for ~22% (0.56 EJ) of the bioenergy produced in 2011 (EIA, 2012b).

In stand-alone facilities in the United States, production of electricity from wood and waste biomass occurs primarily in electricity-only plants (Aguilar et al., 2011). In 2008, only one-fifth of the electricity generated from biomass originated from CHP facilities (White, 2010). Limited expansion of biomass CHP is expected in the United States because of the availability of low-cost fuels, such as domestic natural gas (NAFO, 2013).

Canada

Canada is the only country in this comparison where biomass is not the largest source of renewable energy. In 2011, hydro-electricity accounted for 71% of the renewable energy production in the country. That same year, biomass accounted for just over 25% of the renewable energy production, or ~4.4% (~0.52 EJ) of the total primary energy demand (NEB, 2014).

The contribution of bioenergy to the total energy mix in Canada is much smaller than in Finland, Sweden or the United States. As mentioned earlier, bioenergy accounted for only ~0.52 EJ of the renewable energy produced in Canada in 2011: ~100 PJ were derived from combustion in residential heating applications; ~400 PJ were derived from industrial use of wood waste and spent pulping liquors; and the remaining ~16 PJ were derived from heat and power in DHS and CHP facilities. In fact, bioenergy generation has declined steadily in Canada, from 0.58 EJ in 2007 to 0.52 EJ in 2011, largely because of pulp mill closures stemming from the poor economic climate and increased competition from pulp producers in the Southern Hemisphere. However, the pulp and paper sector continues to be the single largest contributor to Canadian bioenergy, accounting for >50% of the industrial use of biomass (NEB, 2014). Despite the mill closures, 39 facilities across the country are currently operating on-site co-generation facilities.

Biomass for DHS and CHP production in stand-alone facilities, though limited, has enormous potential in Canada. In 2012, installed capacity of co-generation facilities in Canada was 466 MW_{el} and 20 MW_{th}, while installed capacity of heat-only facilities was 75.5 MW_{th}. Since 2000, DHS have undergone a rapid expansion in small communities, particularly in Quebec and British Columbia, and represent the greatest growth in the Canadian bioenergy sector (Bradley and Bradburn, 2012). By 2012, 12 more DHS were under construction, with a total installed capacity of 43.8 MW_{th}.

Since the early 2000s, Canada has emerged as an important producer of wood pellets to supply the growing global market. Installed capacity for wood pellet production increased from ~540,000 tonnes in 2003 to 3.22 million tonnes in 2011; over 50% of this expansion occurred in British Columbia (Bradley and Bradburn, 2012). Not all facilities are operating at full capacity, due to operational and market constraints. In 2011, wood pellet production was estimated at 1.75 million tonnes. Almost all of the pellets produced in Canada are exported (~1.25 million tonnes) because the domestic market for pellets is constrained by the availability of low-cost conventional fossil fuels and hydro-electricity (see the discussion on factors determining bioenergy deployment in chapter: Constraints and Success Factors for Woody Biomass Energy Systems in Two Countries With Minimal Bioenergy Sectors). Europe is the primary market for Canadian pellets, accounting for ~90% of exports, although the recent development of large wood pellet facilities in the US Southeast has begun to corrode Canada's market share. Canadian pellet producers are looking to Asia in the hope that new renewable energy targets in South Korea and Japan may create a favourable market for pellets in the region (Bradley and Bradburn, 2012) (see chapter: Challenges and Opportunities for International Trade in Forest Biomass for a discussion on trade).

CHALLENGES AND OPPORTUNITIES FOR FOREST BIOMASS BASED ENERGY CONVERSION TECHNOLOGIES

As global demand for bioenergy increases, a number of challenges are emerging that affect every aspect of forest bioenergy. These challenges can be broadly grouped as technological-, market- or policy-related.

Technological Challenges

In reviewing biomass upgrading and conversion technologies, it is apparent that significant technological challenges are limiting the commercialisation of new bioenergy technologies. A host of different technologies are available both to upgrade forest biomass into products with improved characteristics and to enhance the efficiency of conversion from forest biomass into energy. However, these technologies remain in the development or pre-commercial stage. High processing costs are recognised as the single largest impediment to commercialisation for most of these technologies. In order for full commercialisation to be realised, further research and development is required to bring down costs.

Market Challenges

Bioenergy also experiences significant difficulties in accessing energy markets worldwide. Three prominent challenges are slowing the integration of bioenergy into the market: the existence of low-cost alternatives, the high capital costs associated with required infrastructure and equipment upgrades and competition for biomass between different world markets. In Canada and the United States, heat and electricity costs are some of the lowest in the developed world due to the abundance of hydro-electric power stations in the provinces of British Columbia and Quebec, and large domestic oil and gas reserves in Alberta, Texas, North Dakota and Montana. In addition, Canadian and American cities lack the DHS infrastructure that has allowed co-gen facilities in Scandinavia to improve their conversion efficiencies and cost-competitiveness. As well, the transition from burning coal to burning biomass requires costly infrastructure upgrades. Biomass resources are consumed for a variety of products, in a number of sectors of the global economy. Bioenergy feedstocks derived from forest biomass will be subject to competition between these differing end-uses. This competition may impede the expansion of forest bioenergy as, in the absence of government policy supporting renewable technologies, bioenergy is of lower value than other forest products, such as lumber and pulp. Thus, bioenergy often struggles to penetrate some energy markets.

Policy Challenges

As highlighted earlier, the presence and structure of government policies have an enormous influence on the success of bioenergy in different regions. In

Canada, the majority of pellets are produced from forest residues. However, the use of roundwood for pellet production is increasing. Current forest harvest policies restrict access to biomass residues, because large quantities (~12% of total harvest volume) of slash are left on the harvest site and sometimes burned (Mabee and Saddler, 2010). A revision of these practices could make available an enormous volume of forest biomass feedstock, with great market potential for bioenergy applications, while allowing sufficient biomass to be left on-site for nutrient recycling. Harvesting of these residues will, however, add significant costs (Ralevic et al., 2010). Additionally, some regions are plagued by a lack of clear, long-term government policies that support bioenergy. In Canada and the United States, this has led to subdued investor confidence and limited commercialisation of large-scale bioenergy projects.

It is also worth noting that different drivers for forest-based bioenergy in Canadian, American and European governments create variable levels and structures of support. In Canada, with government policy primarily focused on enhancing forest sector innovation, government programmes have been structured to stimulate a wide array of projects including bioenergy, biomaterials and biochemicals. Thus, bioenergy projects compete with a host of other technologies for available funding. Due to the relatively low market value of energy products (especially in Canada, where energy is cheap and abundant), bioenergy projects struggle to receive funding. In the United States, the desire for improved energy security has placed the development of biofuels at the top of the funding agenda. In Europe, the desire to mitigate climate change by reducing energy-related carbon emissions has favoured investment in bioenergy technologies, rather than biofuels. In the European context, bioenergy is widely available, can be produced using cost-competitive technologies and has been used to replace emissions-intensive coal-fired power plants and residential oil furnaces. However, as demand for forest-based bioenergy increases, it will become more difficult to ensure the sustainability of biomass feedstocks. Forest plantations established specifically for bioenergy production are likely to become more common under a high bioenergy-use scenario. To maintain GHG emissions reductions, these plantations must adhere to sustainable management practices, or risk increasing the GHG emissions profile of bioenergy projects (Schulze et al., 2012). Maintaining low GHG emissions throughout the life cycle of forest biomass-based energy will be especially important in the EU. That being said, no specific sustainability criteria for the use of solid biomass will be established by the EU until 2020, although five member states have imposed national sustainability criteria (Belgium, Hungary, Italy, the Netherlands and the United Kingdom).

Opportunities for Forest Biomass-Based Energy and Conversion Technologies

Despite the challenges associated with forest biomass-based energy projects, bioenergy is expected to remain the largest global source of renewable energy through to 2035 (IEA, 2013a). Existing and emerging biomass conversion technologies offer enormous potential in both developing and developed countries. Deployment of these technologies can improve the efficiency of biomass conversion, aid rural development, reduce energy-related carbon emissions and enhance energy security.

Improvement in Conversion Efficiency

In developing countries, the transition from traditional, low-grade heating and cooking to more modern bioenergy applications is likely to significantly improve energy efficiency and reduce biomass demand. This will be increasingly important in rapidly developing countries such as Brazil, Russia, India and China, where energy intensity will increase with economic development. That being said, the uptake of technologies such as modern cooking stoves may face obstacles from consumers. Emerging bioenergy technologies can also be used to improve the efficiency of biomass conversion in existing modern applications. Uptake of CHP facilities, instead of heat or power-only facilities, will drastically improve overall efficiency.

Rural Development

Since the global recession began in 2008, rural communities in developed countries around the world have experienced difficult economic climates. These regions, traditionally defined by agriculture, forestry and mining, saw disproportionate job losses as global demand for resources declined. The North American forest sector was particularly affected and the poor economic climate, combined with intensifying competition from mills in the Southern Hemisphere, led to a number of mill closures. Production of bioenergy and associated products (eg wood pellets) offers rural communities a new source of income and has the ability to leverage existing, underutilised capital.

Emissions Reduction and Enhanced Energy Security

Deployment of bioenergy is also capable of enhancing energy security, while effectively reducing energy-related carbon emissions. Unlike deposits of fossil fuels, forest biomass is distributed globally. Bioenergy production can leverage this domestic energy source to improve energy security in regions that have been reliant on oil imports. Deployment of bioenergy technologies in the Nordic countries has successfully reduced their reliance on fossil fuels, thereby improving national energy security. If forest biomass is harvested in a sustainable manner, it can provide, over time, benefits in terms of reduction of carbon emissions to the atmosphere relative to fossil fuels (see the discussion in chapter: Environmental Sustainability Aspects of Forest Biomass Mobilisation). Thus, forest biomass can effectively reduce energy-related carbon emissions when used to replace fossil fuels as an energy source (IEA, 2013a).

SUMMARY AND CONCLUSIONS

The majority of forest biomass is used for energy generation, with 60% employed for inefficient, traditional combustion. Modern bioenergy is also dominated by combustion, where biomass is converted to energy in highly efficient boilers, furnaces and stoves. Modern combustion technologies are used to generate electricity and heat and are used in pulp and paper mills, power generation and DHS. Although industry, particularly the forest products industry, makes good use of bioenergy, this is not expected to increase dramatically in the future. By contrast, the use of biomass to produce domestic power and transportation fuels is expected to increase significantly. To meet the growing demand, biomass mobilisation and supply chains must be optimised and industrial processes for the conversion of biomass into alternative biomass energy products must be commercialised. Pyrolysis, gasification and biochemical conversion technologies offer the opportunity to create more adaptable and energy-dense products. In the same way that wood pellets have enhanced the global mobilisation of biomass, upgrading technologies provide an opportunity to further enhance forest biomass mobilisation. Additionally, these technologies will be required for the development of transportation fuels and, although they have yet to reach full-scale commercialisation, new biofuel/biorefinery applications will likely contribute to future global biomass demand.

Because of their relatively low density, raw forest biomass feedstocks are best suited to local consumption. However, biomass supply and biomass demand do not always occur in the same jurisdictions. Pelletisation is a rapidly growing industry because it increases the density of forest biomass and improves its suitability for transportation and utilisation in jurisdictions without an abundant supply of biomass. Nevertheless, global biomass/biofuels trade volumes remain a small fraction of global energy production and use (including fossil fuels). Where bioenergy demand is expected to increase, biomass upgrading technologies will likely gain increasing prominence. Although Finland, Sweden, the United States and Canada all have an abundant supply of forest biomass, there are distinct variations in the drivers influencing forest biomass-based energy production in each country. Differences in forest ownership influence biomass allocation and use, but differences in priorities, such as improved energy security, climate change mitigation and restructuring of the forest industry, also play important and varying roles in each country. As a consequence, government programmes that support bioenergy also vary, depending on national priorities.

Despite numerous challenges, bioenergy is expected to remain the world's largest form of renewable energy until at least 2035. Existing and emerging technologies offer enormous opportunities to increase the efficiency of biomass conversion, aid in rural development, reduce energy-related carbon emissions and improve national energy security.

REFERENCES

Acharya, B., Sule, I., Dutta, A., 2012. A review on advances of torrefaction technologies for biomass processing. Biomass Conv. Biorefin. 2 (4), 349–369.

Aguilar, F.X., Song, N., Shifley, S., 2011. Review of consumption trends and public policies promoting woody biomass as an energy feedstock in the US. Biomass Bioenergy 35 (8), 3708–3718.

Ahrenfeldt, J., Thomsen, T.P., Henriksen, U., Clausen, L.R., 2013. Biomass gasification cogeneration—a review of state of the art technology and near future perspectives. Appl. Therm. Eng. 50 (2), 1407–1417.

Al-Mansour, F., Zuwala, J., 2010. An evaluation of biomass co-firing in Europe. Biomass Bioenergy 34 (5), 620–629.

Arbogast, S., Bellman, D., Paynter, J., Wykowski, J., 2013. Advanced biofuels from pyrolysis oil: opportunities for cost reduction. Fuel Proc. Technol. 106, 518–525.

Arena, U., 2012. Process and technological aspects of municipal solid waste gasification. A review. Waste Manag. 32 (4), 625–639.

Aslani, A., Naaranoja, M., Helo, P., Antila, E., Hiltunen, E., 2013. Energy diversification in Finland: achievements and potential of renewable energy development. Int. J. Sustain. Energy 32 (5), 504–514.

Balat, M., Balat, M., Kırtay, E., Balat, H., 2009. Main routes for the thermo-conversion of biomass into fuels and chemicals. Part 2: Gasification systems. Energy Conv. Manag. 50 (12), 3158–3168.

Batidzirai, B., Mignot, A., Schakel, W., Junginger, H., Faaij, A., 2013. Biomass torrefaction technology: techno-economic status and future prospects. Energy 62, 196–214.

Bauen, A., Berndes, G., Junginger, M., Londo, M., Vuille, F., Ball, R., Bole, T., Chudziak, C., Faaij, A., Mozaffarian, H., 2009. Bioenergy—a sustainable and reliable energy source. A review of status and prospects. IEA Bioenergy (2009:06), 108 p. Available from: http://www.ieabioenergy.com/wp-content/uploads/2013/10/MAIN-REPORT-Bioenergy-a-sustainable-and-reliable-energysource.-A-review-of-status-and-prospects.pdf.

Berguson, B., Eaton, J., Stanton, B. Development of hybrid poplar for commercial production in the United States: the Pacific Northwest and Minnesota experience. Sustainable Alternative Fuel Feedstock Opportunities, Challenges and Roadmaps for Six US Regions. Proceedings of the Sustainable Feedstocks for Advance Biofuels Workshop: Soil and Water Conservations Society, 282–299.

Bradley, D., Bradburn, K., 2012. Economic Impact of Bioenergy in Canada—2011. Canadian Bioenergy Association, Ottawa, ON, Available from: http://www.canbio.ca/upload/documents/canbio-bioenergy-data-study-2011-jan-31a-2012.pdf.

Bridgwater, A.V., 2012. Review of fast pyrolysis of biomass and product upgrading. Biomass Bioenergy 38, 68–94.

Butler, E., Devlin, G., Meier, D., McDonnell, K., 2011. A review of recent laboratory research and commercial developments in fast pyrolysis and upgrading. Renew. Sustain. Energy Rev. 15 (8), 4171–4186.

Chew, J.J., Doshi, V., 2011. Recent advances in biomass pretreatment–torrefaction fundamentals and technology. Renew. Sustain. Energy Rev. 15 (8), 4212–4222.

Chum, H., Faaij, A., Moreira, J., Berndes, G., Dhamija, P., Dong, H., Gabrielle, B., Goss Eng, A., Lucht, W., Mapako, M., Masera Cerutti, O., McIntyre, T., Minowa, T., Pingoud, K., 2011. Bioenergy. In: Edenhofer, O., Pichs-Madruga, R., Sokona, Y., Seyboth, K., Matschoss, P., Kadner, S., Zwickel, T., Eickemeier, P., Hansen, G., Schlömer, S., Stechow, C.V. (Eds.), IPCC Special Report on Renewable Energy Sources and Climate Change Mitigation. Cambridge University Press, Cambridge, UK and New York, NY.

Chundawat, S.P., Beckham, G.T., Himmel, M.E., Dale, B.E., 2011. Deconstruction of lignocellulosic biomass to fuels and chemicals. Annu. Rev. Chem. Biomol. Eng. 2, 121–145.

Ciolkosz, D., Wallace, R., 2011. A review of torrefaction for bioenergy feedstock production. Biofuels Bioprod. Biorefin. 5 (3), 317–329.

Damartzis, T., Zabaniotou, A., 2011. Thermochemical conversion of biomass to second generation biofuels through integrated process design—a review. Renew. Sustain. Energy Rev. 15 (1), 366–378.

Dornburg, V., van Vuuren, D., van de Ven, G., Langeveld, H., Meeusen, M., Banse, M., van Oorschot, M., Ros, J., van den Born, G.J., Aiking, H., 2010. Bioenergy revisited: key factors in global potentials of bioenergy. Energy Environ. Sci. 3 (3), 258–267.

EIA, 2009. Industrial Biomass Energy Consumption and Net Generation by Industry and Energy Source. EIA, Washington, DC, Available from: http://www.eia.gov/renewable/annual/trends/pdf/table1_8.pdf.

EIA, 2012a. Primary Energy Consumption Estimates by Source, 1949–2011. EIA, Washington, DC, Available from: http://www.eia.gov/totalenergy/data/annual/showtext.cfm?t=ptb0103.

EIA, 2012b. Wood and Biomass Waste Consumption Estimates, 2009. EIA, Washington, DC, Available from: http://www.eia.gov/renewable/data.cfm#biomass.

Ericsson, K., Huttunen, S., Nilsson, L.J., Svenningsson, P., 2004. Bioenergy policy and market development in Finland and Sweden. Energy Policy 32 (15), 1707–1721.

Faaij, A.P., 2006. Bio-energy in Europe: changing technology choices. Energy Policy 34 (3), 322–342.

FAO, 2013. FAO Yearbook Forest Products 2011. FAO, Rome, Italy.

Fernando, R., 2009. Co-Gasification and Indirect Cofiring of Coal and Biomass. IEA Clean Coal Centre, London, UK, PF 09-14, December 2009. Available from: http://www.iea-coal.org.uk/documents/82242/7415/Co-gasification-and-indirect-cofiring-of-coal-and-biomass-(CCC/158).

Goh, C.S., Junginger, M., Cocchi, M., Marchal, D., Thrän, D., Hennig, C., Heinimö, J., Nikolaisen, L., Schouwenberg, P.-P., Bradley, D., Hess, R., Jacobson, J., Ovard, L., Deutmeyer, M., 2013. Wood pellet market and trade: a global perspective. Biofuels Bioprod. Biorefin. 7 (1), 24–42.

Göransson, K., Söderlind, U., He, J., Zhang, W., 2011. Review of syngas production via biomass DFBGs. Renew. Sustain. Energy Rev. 15 (1), 482–492.

Grahn, M., Hansson, J., 2015. Prospects for domestic biofuels for transport in Sweden 2030 based on current production and future plans. WIREs Energy Environ. 4, 290–306.

Haberl, H., Beringer, T., Bhattacharya, S.C., Erb, K.-H., Hoogwijk, M., 2010. The global technical potential of bio-energy in 2050 considering sustainability constraints. Curr. Opin. Environ. Sustain. 2 (5), 394–403.

Hakkila, P., 2006. Factors driving the development of forest energy in Finland. Biomass Bioenergy 30 (4), 281–288.

Hoogwijk, M., Faaij, A., van den Broek, R., Berndes, G., Gielen, D., Turkenburg, W., 2003. Exploration of the ranges of the global potential of biomass for energy. Biomass Bioenergy 25 (2), 119–133.

IEA, 2012a. Technology Roadmap: Bioenergy for Heat and Power. IEA, Paris, France.

IEA, 2012b. World Energy Outlook 2012. IEA, Paris, France.

IEA, 2013a. World Energy Outlook 2013. IEA, Paris, France.

IEA, 2013b. Nordic Energy Technology Perspectives. IEA, Paris, France.

Jørgensen, H., Kristensen, J.B., Felby, C., 2007. Enzymatic conversion of lignocellulose into fermentable sugars: challenges and opportunities. Biofuels Bioprod. Biorefin. 1 (2), 119–134.

Karatzos, S., McMillan, J.D., Saddler, J.N., 2014. The potential and challenges of drop-in biofuels. Report T39-T1. IEA Bioenergy. Available from: http://task39.org/files/2014/01/Task-39-drop-in-biofuels-report-summary-FINAL-14-July-2014-ecopy.pdf.

Kirkels, A.F., Verbong, G.P., 2011. Biomass gasification: still promising? A 30-year global overview. Renew. Sustain. Energy Rev. 15 (1), 471–481.

Langeveld, H., Quist-Wessel, F., Dimitriou, I., Aronsson, P., Baum, C., Schulz, U., Bolte, A., Baum, S., Köhn, J., Weih, M., 2012. Assessing environmental impacts of short rotation coppice (SRC) expansion: model definition and preliminary results. Bioenergy Res. 5 (3), 621–635.

Li, H., Liu, X., Legros, R., Bi, X.T., Lim, C.J., Sokhansanj, S., 2012. Pelletization of torrefied sawdust and properties of torrefied pellets. Appl. Energy 93, 680–685.

Lipp, J., 2007. Lessons for effective renewable electricity policy from Denmark, Germany and the United Kingdom. Energy Policy 35 (11), 5481–5495.

Mabee, W.E., Mirck, J., 2011. A regional evaluation of potential bioenergy production pathways in Eastern Ontario, Canada. Ann. Assoc. Am. Geogr. 101 (4), 897–906.

Mabee, W., Saddler, J., 2010. Bioethanol from lignocellulosics: status and perspectives in Canada. Biores. Technol. 101 (13), 4806–4813.

Menon, V., Rao, M., 2012. Trends in bioconversion of lignocellulose: biofuels, platform chemicals and biorefinery concept. Prog. Energy Combust. Sci. 38 (4), 522–550.

Mohan, D., Pittman, C.U., Steele, P.H., 2006. Pyrolysis of wood/biomass for bio-oil: a critical review. Energy Fuels 20 (3), 848–889.

NAFO, 2013. Update and Context for U.S. Wood Bioenergy Markets, Washington, DC. Forisk Consulting, Athens, GA, Available from: http://www.theusipa.org/Documents/NAFO-US_Bioenergy_Markets-FINAL-201306261.PDF.

NEB, 2014. Canadian Energy Overview 2013—Energy Briefing Note. National Energy Board Government of Canada, Calgary, AB, Available from: https://www.neb-one.gc.ca/nrg/ntgrtd/mrkt/vrvw/2013/index-eng.html.

Nicholls, D., Monserud, R., Dykstra, D., 2008. A Synthesis of biomass utilization for bioenergy production in the Western United States, General Technical Report-Pacific Northwest Research Station. USDA Forest Service (PNW-GTR-753).

Pereira, E.G., da Silva, J.N., de Oliveira, J.L., Machado, C.S., 2012. Sustainable energy: a review of gasification technologies. Renew. Sustain. Energy Rev. 16 (7), 4753–4762.

Ralevic, P., Ryans, M., Cormier, D., 2010. Assessing forest biomass for bioenergy: operational challenges and cost considerations. Forestry Chron. 86 (1), 43–50.

Röser, D., Asikainen, A., Stupak, I., Pasanen, K., 2008. Forest energy resources and potentials. In: Röser, D., Asikainen, A., Raulund-Rasmussen, K., Stupak, I. (Eds.), Sustainable Use of Forest Biomass for Energy. Springer, Dordrecht, the Netherlands, pp. 9–28.

Salomón, M., Savola, T., Martin, A., Fogelholm, C.-J., Fransson, T., 2011. Small-scale biomass CHP plants in Sweden and Finland. Renew. Sustain. Energy Rev. 15 (9), 4451–4465.

Schmidt, J., Gass, V., Schmid, E., 2011. Land use changes, greenhouse gas emissions and fossil fuel substitution of biofuels compared to bioelectricity production for electric cars in Austria. Biomass Bioenergy 35 (9), 4060–4074.

Schulze, E.D., Körner, C., Law, B.E., Haberl, H., Luyssaert, S., 2012. Large-scale bioenergy from additional harvest of forest biomass is neither sustainable nor greenhouse gas neutral. GCB Bioenergy 4 (6), 611–616.

SEA, 2013. Energy in Sweden 2012 Swedish Energy Agency. Available from: http://www.energimyndigheten.se/Global/Engelska/Factsandfigures/Energy_in_sweden_2012.pdf

Sikkema, R., Steiner, M., Junginger, M., Hiegl, W., Hansen, M.T., Faaij, A., 2011. The European wood pellet markets: current status and prospects for 2020. Biofuels Bioprod. Biorefin. 5 (3), 250–278.

Sixta, H., 2006. Introduction. Handbook of Pulp. Wiley-VCH Verlag GmbH, Weinheim, Germany, pp. 2–19.

Statistics-Finland, 2014. Statistics Finland—Energy. Available from: http://www.stat.fi/til/ene_en.html

Stelte, W., Sanadi, A.R., Shang, L., Holm, J.K., Ahrenfeldt, J., Henriksen, U.B., 2012. Recent developments in biomass pelletization—a review. BioResources 7 (3), 4451–4490.

SVEBIO, 2013. Swedish Bioenergy Association. Bioenergy Facts. Available from: http://www.svebio.se/english/bioenergy-facts

Tumuluru, J.S., Wright, C.T., Hess, J.R., Kenney, K.L., 2011. A review of biomass densification systems to develop uniform feedstock commodities for bioenergy application. Biofuels Bioprod. Biorefin. 5 (6), 683–707.

UNICA, 2014. Production Data. Available from: http://www.unicadata.com.br/index.php?idioma=2

Vakkilainen, E., Kuparinen, K., Heinimö, J., 2013. Large Industrial Users of Energy Biomass. IEA Bioenergy Task 40, Lappeenranta, Finland, Available from: http://www.bioenergytrade.org/downloads/t40-large-industrial-biomass-users.pdf.

Westholm, E., Lindahl, K.B., 2012. The Nordic welfare model providing energy transition? A political geography approach to the EU RES directive. Energy Policy 50, 328–335.

White, E., 2010. Woody biomass for bioenergy and biofuels in the United States—a briefing paper, General Technical Report-Pacific Northwest Research Station, USDA Forest Service (PNW-GTR-825).

Chapter 8

Challenges and Opportunities for International Trade in Forest Biomass

Patrick Lamers*, Thuy Mai-Moulin**, Martin Junginger**
*Idaho National Laboratory, Idaho Falls, ID, United States of America; **Copernicus Institute, Utrecht University, Utrecht, The Netherlands

Highlights

- By 2020, the global demand for internationally traded wood pellets from boreal and temperate forests is expected to reach 15–26 million tonnes (264–458 PJ) per year.
- For the foreseeable future, critical demand markets will remain in the EU and, to a lesser extent, Asia.
- Several key importing countries, including the United Kingdom, the Netherlands, Belgium and Denmark, have developed or are in the process of developing sustainability requirements for woody biomass acquisition and consumption.
- At present, the United States, Canada and Russia are the most important suppliers of traded wood pellets from temperate and boreal biomes and this is not expected to change before 2020.
- Some of the key challenges facing countries from which wood pellets are exported include: limited incentives for small-holder SFM certification across the Southeast United States, cultural differences in forest and land management definitions between Canada and the EU, as well as concerns regarding the efficacy of SFM auditing/monitoring in Northwest Russia.

INTRODUCTION

Fossil fuels currently satisfy approximately 81% of global primary energy needs, with consumption projected to increase further, due to rising incomes and a growing population (IEA, 2012). However, fossil fuel combustion is one of the key sources of greenhouse gas (GHG) emissions and, thus, a major contributor to anthropogenic climate change (IPCC, 2007, 2014). One of the critical challenges for the energy sector this century will be to decouple energy supply from GHG emissions. Bioenergy has been proposed as one option to address this challenge (Creutzig et al., 2014).

In some markets (eg Scandinavia), industrial-scale use of forest biomass as a source of bioenergy has been part of the national energy strategy for several decades (see chapter: Comparison of Forest Biomass Supply Chains From the Boreal and Temperate Biomes). In most markets, however, large-scale modern bioenergy production is relatively recent (REN21, 2014). There are numerous objectives driving the increased use of bioenergy including, but not limited to, reduction of GHG emissions, enhanced energy security and diversification and support for domestic industries (agriculture, forestry, processing etc.). Within a single decade, these factors have contributed to a shift in the way bioenergy is viewed, from a largely domestic resource to a globally traded one (Chum et al., 2011; Lamers, 2014). Indeed, more than 300 PJ of woody biomass were directly traded for energy on the international market in 2010 (Lamers et al., 2012).

Over the past several years, most of the globally traded solid biomass has been used for energy production in the EU (Junginger et al., 2013; Lamers et al., 2014a, 2014b). Other key importing regions for internationally traded forest biomass include South Korea and Japan. Cross-border trade also exists (eg between the United States and Canada, between Norway and Sweden). South Korea, Japan and many EU countries are expected to remain net importers of forest biomass for energy production (Kranzl et al., 2014). Recent analyses suggest that the EU's total woody biomass imports for energy (including forest biomass and agricultural residues) could increase by over 400% between 2010 and 2020 (Lamers et al., 2015).

International trade in woody biomass from temperate and boreal forests could be an important element in the mobilisation of bioenergy. For example, the recent expansion of wood pellet production capacity in the United States and Canada has been directly linked to export market developments in Europe and Asia. At present, renewable energy targets in export markets appear to be among the key drivers for international woody biomass trade (Junginger et al., 2014). Although trade barriers (eg import taxes and duties) are common for liquid biofuels, they have not, to date, been applied to woody biomass on a large scale. The need for internationally recognised technical standards and uniform contracts has also been addressed in recent years. The key remaining requirements for the development of a mature, international-scale woody biomass trade include: development of phytosanitary restrictions (to prevent the spread of vermin and fungi), reduction in logistical costs and, most importantly, establishment of policy frameworks to ensure sustainable sourcing (Junginger et al., 2014).

Sustainability criteria for forest biomass may influence the type and volume of feedstocks available to producers (eg wood pellet companies) and traders exporting to high-demand markets. In other words, sustainability standards for forest biomass designated for international trade could pose challenges, but also opportunities, for all parties involved in the forest bioenergy market. This chapter aims at defining the key barriers and opportunities for international forest bio-energy trading, with a special focus on those that may arise due to sustainability requirements within existing or proposed legislative frameworks in key regions. It also sets out to derive quantitative estimates of the demand for and supply of woody biomass from boreal and temperate forests by 2020 and highlight areas where proposed sustainability frameworks could lead to conflicts.

With a focus on wood pellets derived from boreal and temperate forests, this chapter first identifies and quantifies the most important destination markets for globally traded wood pellets. Next, it identifies and quantifies the most important current and expected sources of wood pellets and characterises these regions in terms of their competitiveness (eg via cost of supply, state of the industry). Finally, it provides an overview of the sustainability requirements and initiatives (current and expected) imposed by key destination markets and the sustainability requirements and certification status (current and expected) of the major wood pellet source markets; it also discusses their implications for international trade.

DEMAND MARKETS

European Union

As laid out in RED 2009/28/EC, 20% of final energy consumption must be provided by renewable energy across all member states within the EU, by 2020. Individual member states' National Renewable Energy Action Plans (NREAP) describe technological and sector trajectories and policy frameworks that have been put in place to achieve this goal. According to the NREAPs, approximately 42% of the total renewable energy target must be achieved via combustion of biomass for electricity, heating and cooling by 2020. Most of this energy will be used for heating/cooling in the residential sector and come from solid biomass (AEBIOM, 2012). Until now, the vast majority of the EU's woody biomass demand has been supplied domestically. However, the EU has attracted most of the international trade in forest biomass over the past decade (Lamers et al., 2012). Wood chips, waste wood and roundwood have been imported from bordering countries, while wood pellets have been traded cross-continentally, particularly from North America and Russia (Figs 8.1 and 8.2).

The most important EU market for wood pellet imports is the industrial sector, that is large-scale (>5 MW$_{el}$) co-firing and dedicated heat and power installations. Because of anticipated increases in demand under current policy projections and inadequate regional resources (eg limited available land, high feedstock costs and unacceptable delays in mobilisation), the United Kingdom, the Netherlands, Belgium, Denmark and Sweden are expected to remain net importers of woody biomass until at least 2020 (Kranzl et al., 2014).

It is important to note that Sweden and Denmark source wood pellets primarily from countries around the Baltic Sea, including Estonia, Latvia, Russia and Germany (Fig. 8.2). Denmark has also imported wood pellets from Portugal.

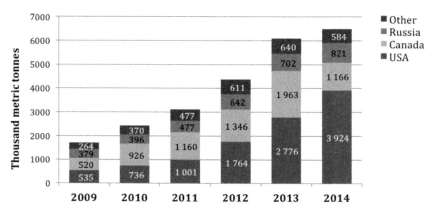

FIGURE 8.1 **Extra-EU imports of wood pellets to EU member states.** Note: Croatia became the 28th member state of the EU in 2014.

By contrast, the United Kingdom, the Netherlands and Belgium import wood pellets predominantly from the United States and Canada.

For the most part, the international trade of wood pellets for industrial use is negotiated directly between buyer and seller. Although there is generally little re-distribution of internationally traded wood pellets among European countries, trade streams (eg from the Netherlands to Germany or Denmark) have existed in the past (Fig. 8.2). Within the EU, there are strong differences in the patterns of trade in wood pellets for residential and industrial applications. Latvia, Estonia, Portugal and Finland are the largest exporting member states of industrial-use wood pellets (Table 8.1). Germany, Austria and the Balkan States have been the most important sources of wood pellets for Italy, which is the largest importer of pellets, predominantly for residential use. Denmark is the key importing member state of industrial-use wood pellets in the EU (Table 8.1), but Belgium, the Netherlands and the United Kingdom are also major importers of industrial-use forest biomass from outside the EU (Goh et al., 2013; Lamers et al., 2012). The demand for tradable solid biomass is expected to increase significantly in all of these countries because of proposed and/or existing policy measures that encourage the installation of large-scale co- and mono-firing plants for the generation of electricity (and heat) using solid biomass (Beurskens and Hekkenberg, 2010; Sikkema et al., 2011a).

Belgium

Belgium's current installed capacity of 280 MW_{el} per year is estimated to reach 730–900 MW_{el}/year (equivalent to approximately 3 million tonnes of wood pellets) by 2020. However, industry projections (Table 8.2) indicate that Belgium is unlikely to reach half of its initially projected NREAP level (2 GW_{el}). Current policy discussions indicate that subsidies from the national government will only be granted to power plants that provide full disclosure of their capital and operational expenditures.

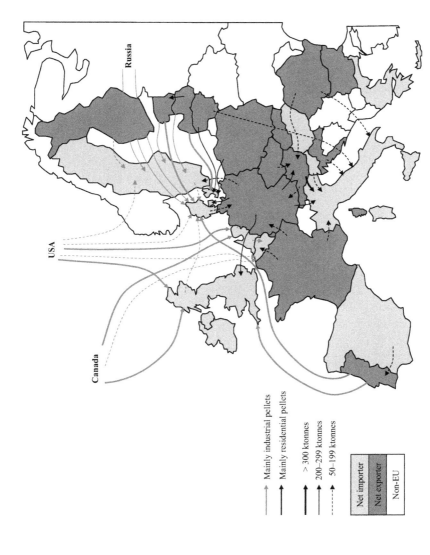

FIGURE 8.2 Annual wood pellet trade flows across the EU between 2010 and 2013.

TABLE 8.1 Intra-EU Trade in Wood Pellets (in Ktonnes)

Country	2012		2013		Market share		Dominant wood pellet type per application
	Imports	Exports	Imports	Exports	Imports (%)	Exports (%)	
Germany	265	515	273	517	6	11	Residential
Estonia	131	402	114	677	3	12	Industrial
Latvia	3	760	7	1216	0	22	Industrial
Austria	323	361	192	380	6	8	Residential
Portugal	2	371	9	474	0	9	Industrial
Finland	15	43	3	61	0	1	Industrial
Romania	1	212	1	355	0	6	Residential
Spain	17	38	15	74	0	1	Residential
Poland	91	84	35	192	1	3	Residential
Denmark	1434	140	1322	128	31	3	Industrial
Lithuania	15	165	41	184	1	4	Residential
Netherlands	305	100	245	118	6	2	Industrial
Sweden	351	148	375	126	8	3	Industrial
United Kingdom	296	129	304	117	7	3	Industrial
Belgium	288	21	174	103	5	1	Industrial
Italy	1034	3	957	19	22	0	Residential
Other	217	346	122	429			
EU	4790	3838	4190	5170			

TABLE 8.2 Expected Biomass Cofiring Capacity Developments in Belgium

Company	Plant	Completion date	Capacity (MW$_{el}$)	Biomass demand (Mtonnes)
Electrabel	Max Green (phase-out)	–	(220)	
E.ON	Langerlo (conversion)	2015	400	1.25
BEE	Gent (new plant)	2017	180–200	0.75
E.ON	Antwerp (new plant)	2018	150–300	0.5–1
SUM			730–900	2.5–3

Denmark

Recent projections conducted by the Danish government suggest that wood pellet consumption will increase to approximately 50 PJ (heat and power sector) and 10 PJ (industrial sector) by 2020 (DEA, 2012a). Danish energy supplier Dong recently announced the conversion of two additional plants (Aalborg and Aarhus) to wood pellet use; however, the conversion of its Studstrup plant has been delayed until 2016 (Dale, 2013). In recent years, Denmark has imported between 1–2 million tonnes of wood pellet per year and estimates suggest that annual imports of wood pellets by the industrial sector could reach close to 3 million tonnes, by 2020 (Table 8.4).

The Netherlands

The Dutch Energy Accord (published in 2013) has set a maximum target of 25 PJ output, to be generated via co-firing in existing (Amer and Hemweg) and newer, more efficient power plants, by 2020. Older coal-fired power plants from the 1980s are slated for shut down by 2017. On a biomass basis, 25 PJ of electricity would equal close to 7 TWh, or 3.5 million tonnes of wood pellets. While it is not clear how industry will meet this target, we estimate a biomass demand of 3.5 million tonnes by 2020 (Table 8.4), assuming an efficiency factor for co-firing of 43% (using data for both old and new coal plants). Because funds from the Dutch Renewable Energy Subsidy Scheme (SDE +) have already been fully allocated for other renewable energy options until 2017, it is expected that large-scale co-firing will commence again in 2018.

Sweden

Woody biomass co- and mono-firing combined heat and power (CHP) installations in Sweden have traditionally sourced large shares of their demand from within Sweden and from neighbouring Norway and Finland. However, the price and availability of woody biomass feedstocks (eg the tops and branches generated during forest harvesting) are dependent on the timber, pulp and paper markets in those countries. Imports to Swedish CHP plants (which are considerably smaller than co-firing installations in the United Kingdom and the Netherlands) have been on ad hoc basis and in smaller volumes because of a preference for local suppliers and because of shipping size limitations for the Baltic Sea. Lately, imports of woody biomass for energy production have been reduced to a few large contracts from Russia, while smaller, ad hoc contracts have been cancelled (Hektor, B., 2014. RE: Member Representative Swedish Bioenergy Association SVEBIO, Stockholm, Sweden, Personal Communication). The Swedish pellet market is currently declining because most district heating companies have switched to wood chips (or municipal solid waste) to satisfy demand base loads and only use pellet boilers for peak loads (Olsson, O., 2014. RE: Stockholm Environment Institute, Personal Communication). These wood chips originate from raw forest material and from demolition/waste wood. In the past, Sweden has been a key importer of demolition/waste wood from within Europe, due to the market advantage of flue gas cleaning technologies (Lamers et al., 2012). However, demand for this material from other markets (especially in the United Kingdom) continues to increase. Thus, it is unlikely that Sweden will remain a significant importer of woody biomass for the industrial heating and electricity sector.

The United Kingdom

The current and anticipated UK policy framework is expected to trigger several new investments in electricity generation (Table 8.3). By 2020, up to 5130 MW_{el} of new mono- and/or co-firing capacity for biomass utilisation could come online (Blair, 2013). However, the UK support scheme is capped at 5 GW_{el} total biomass co-firing, with a 400 MW_{el} cap for installations without CHP generation (Harrabin, 2013). According to these requirements, only 3930 MW_{el} of the announced investments would qualify for subsidies and are, therefore, expected to utilise biomass. In a recent assessment of the country's electricity generation capabilities, the UK's Office of Gas and Electricity Markets calculated a biomass capacity of 3895 MW_{el} by 2018/2019 (OFGEM, 2013) (Table 8.4).

In Table 8.4, industry projections for each of the five countries described earlier are compared with policy projections made via the member states' respective NREAPs. For each country, the grid-connected heat production volume, that is the fraction of new capacity for biomass-based electricity production generated by CHP, was calculated using the relationship between electrical and grid-connected heat generation in the respective member states' NREAPs. The demand estimates outlined in Table 8.4 were employed for scenario analyses in Lamers et al. (2015), which focused on Northwest Europe and correspond to assumptions/scenario analyses by Pöyry (2014).

TABLE 8.3 Expected Biomass Cofiring Capacity Developments in the United Kingdom

Company	Plant	Completion date	Capacity (MW$_{el}$)	Biomass demand (Mtonnes)
RWE	Tilbury, 2 units	2011	(Offline) 750	(Offline) 2–3
E.ON	Ironbridge, 2 units	2013	(Offline) 900	(Offline) 2–3
Drax	Drax, 3 of 6 units	2013, 2014, 2017	3 × 600	6–8
Eggborough Power	Eggborough, 4 units	2015	4 × 500	6–8
RWE	Lynemouth, 3 units	n/a	3 × 110	1
International Power	Rugeley, 2 units	n/a	2 × 500	2–4
TOTAL (theoretical)			5130	15–21
of which supported			3930	

Asia

China is expected to significantly increase its utilisation of biomass for the production of power and heat (Roos and Brackley, 2012), albeit exclusively from local agricultural and forest residues (Cocchi et al., 2011; Pöyry, 2011). Thus, it is not expected to become a strong competitor for internationally traded woody biomass. By contrast, international imports of tradable woody biomass are likely to expand in Japan and South Korea, where new policies are expected to increase the local demand for large- (Japan) and small-scale (South Korea) use. By 2020, Japan and South Korea are expected to consume a combined 22.66 million tonnes (399 PJ) of woody biomass, 15.65 million tonnes (275 PJ) of which could be supplied domestically. This leaves both countries with a combined supply gap of 7 million tonnes by 2020 (Japan: 3 million tonnes; South Korea: 4 million tonnes). Although both countries source globally, a large share of their import volumes have typically originated from within Asia (eg China, Vietnam, Malaysia) (Cocchi et al., 2011) and this is not expected to change. Unfortunately, there is little information available on future developments in wood pellet utilisation capacity across Southeast Asia. That being said, Vietnam is projected to export 3 million tonnes of pellets from wood industry processing residues by 2020 (Cocchi et al., 2011) and conservative estimates suggest that South Korea and Japan may, therefore, consume up to 4 million tonnes of wood pellets from Vietnam alone.

TABLE 8.4 RES-E Capacity and Biomass Demand From Dedicated Mono- or Cofiring Installations in Select EU Countries

	Solid biomass installations[a]	Cofiring capacity	NREAP projections (2020)			Industry projections (2020) wood pellet capacity and use		
	MWel (by 2010)	MWel (by 2012)	MWel	GWh[b]	Mtonnes[c]	MWel	GWh[b]	Mtonnes[c]
BE	727	280	2,007	9,575	5.8	910	4,341	2.6
DK[d]	1,168	(996)[e]	2,404	6,345	3.8	1,814	4,788	2.9
NL	992	413–551[f]	2,253	11,975	7.2	1,306	6,942	3.5
SE	3,823[g]	n/a	2,872	16,635	10.0	n/a	n/a	n/a
UK	2,097	208–338[h]	3,140	20,590	12.4	3,895	25,541	15.4

[a]Data provided by member states in their 2010 status reports to the European Commission (EC) for all solid biomass power installations (excluding biogas and bioliquid installations).
[b]Gross electricity generation.
[c]Biomass demand; assumed energy content (LHV): 17.6 GJ/tonne.
[d]Total installed capacity for solid biomass of all sizes (excluding biogas and bioliquid installations).
[e]DEA (2012b), total installed capacity for solid biomass of all sizes.
[f]Agentschap-NL (2013); the lower value is large-scale installations only, the higher value represents the total installed capacity (ie installations of all sizes).
[g]Includes all municipal solid waste capacity (although only 50% can be accounted for as biomass).
[h]DECC (2013b); variation between 2011 and 2012 due to partial closure of Tilbury power station (RWE/Essent/npower) after a fire.

North America

Across the United States, the demand for locally produced wood pellets is increasing, but not seriously threatening the supply of internationally traded wood pellets. Most large EU-based energy companies are either involved in the production of raw materials upstream (eg Drax in Mississippi, RWE/Essent/npower in Georgia) or specifically produce pellets for export markets overseas (eg Enviva, Fram Renewables, Enova and German Pellets) and can compete against the sale of pellets to domestic markets. Demand markets in the United States do not yet require sustainability criteria for forest biomass, but there are intense and ongoing discussions between the United States and Europe regarding the development of common standards for forest biomass production (Pinchot-Institute, 2013). The Renewable Fuel Standard (RFS2) and the Low Carbon Fuels Standard (for California) could be used as benchmarks for the development of such standards.

Several coal power plants in Ontario, Canada, have recently been converted for wood pellet use, suggesting that Canada could have a local pellet demand for up to 5 million tonnes, by 2019 (Dale, 2013). Wood pellet supply for Ontario is expected to come from within the province, although one station (Thunder Bay) has been converted to consume Steam Explosion Pellets (SEP)—otherwise known as black pellets or bio-coal, imported from Norway as of 2015. The main exporting regions for wood pellets can be found along the west and east coasts of the country (in the provinces of British Columbia, Quebec, New Brunswick and Nova Scotia). Given the long shipping distances and the limited number of water channels linking inland areas to deep-sea harbours, wood pellets currently designated for domestic use are not to become designated for export directly. However, the establishment of new routes for transporting pellets (eg from eastern Ontario to Quebec City by rail and then to the United Kingdom by boat) suggest that this is likely to change.

Global Demand by 2020

The global trade in woody biomass from boreal and temperate forests is driven largely by policy targets and supply costs. The total global demand for wood pellets for heat and power production is expected to reach 32–36 million tonnes by 2020, approximately double the demand in 2010 (Fig. 8.3). Estimates for Northwest Europe (Table 8.4) and calculations by Pöyry (2014) suggest that even high-demand regions, such as the EU, could supply sufficient biomass to meet domestic needs (Lamers et al., 2015). However, internationally traded woody biomass is often cheaper and, thus, preferred over more expensive locally produced biomass. That being said, the United States and Canada are expected to satisfy domestic regional demands due to low supply costs. Japan and South Korea are projected to import 4 million tonnes by 2020. Other regions of the world may require as much as 2 million tonnes of woody biomass per year, but this estimate was not included in Fig. 8.3 because of uncertainty in accounting for respective domestic supplies. In fact, the domestic supply of woody biomass

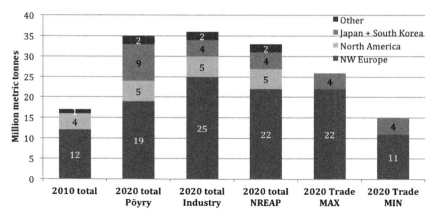

FIGURE 8.3 **Estimated global woody biomass demand for large-scale heat and power generation by 2020.** Note: Northwest Europe includes Belgium, Denmark, the Netherlands and the United Kingdom. Estimates of demand for woody biomass in Japan and South Korea only include that obtained from outside SE Asia. Pöyry 2020 is based on the 'Central scenario' in Pöyry (2014).

within the EU is the single largest uncertainty in predicting the demand for tradable woody biomass by 2020. Thus far, Northwest Europe has been predominantly import-oriented. By 2020, however, a larger fraction could be supplied from within the EU (eg the Baltic States). To account for this, Fig. 8.3 shows two estimates: one for 100% (MAX) and one for 50% (MIN) import dependency. The second estimate is in line with trade projections by Lamers et al. (2014a).

SUPPLY REGIONS

Key Regions of Woody Biomass Production up to 2013

As demonstrated earlier, the EU represents the largest demand market for woody biomass. Although the EU is capable of meeting most of its demand for residential wood pellets from domestic sources, it is heavily dependent on imports for industrial heat and electricity production. Without these imports, the EU would not be capable of meeting NREAP targets. The second largest markets for internationally traded woody biomass are Japan and South Korea, which require imported wood pellets for both residential and industrial use. Although most internationally traded wood pellets imported to Asia and Europe have originated in North America and Russia, Japan and South Korea have also obtained wood pellets from within Southeast Asia. There is limited local demand for wood pellets within Russia and North America. Therefore, the recent increase in wood pellet production in these regions (Fig. 8.4) must be driven by the growing demand from foreign markets. China has also ramped up pellet production from forest and agricultural residues in recent years, but consumption is largely domestic and the smaller fraction of traded material has been limited to shipments within Asia.

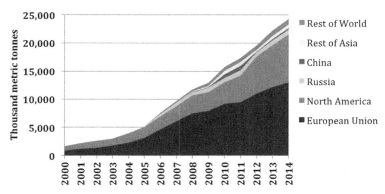

FIGURE 8.4   **Global wood pellet production (AEBIOM, 2013; Goh et al., 2013; Hawkins-Wright, 2013; Lamers et al., 2012, 2014a; REN21, 2015).**

Wood chips and roundwood have also been traded for the production of energy. Most of the trade in wood chips for energy (virgin and/or tertiary waste) is limited to Europe, Turkey and Japan (Lamers et al., 2012). Although the majority of the trade in roundwood is not connected to bioenergy, there is a large amount of indirect trade, in the form of wood processing residues (Heinimö, 2008; Heinimö et al., 2013). The direct trade in raw forest material for bioenergy (ie excluding waste wood) has been small, compared to the trade in wood pellets, because high quality raw wood chips and roundwood can achieve higher prices in the pulp and paper and timber markets (Lamers et al., 2014a). Both wood chips and roundwood are also less suitable for international trade than wood pellets because of their relatively high moisture content and low bulk density, which renders them more expensive to transport over long distances. As a consequence, the trade in wood chips and roundwood for bioenergy have largely been confined to regional and domestic markets.

Key Regions of Woody Biomass Production by 2020

By 2020, the greatest demand for woody biomass from boreal and temperate forests is expected to come from countries/regions with a limited local supply of forest biomass. Recent growth in the production of internationally tradable woody biomass has occurred in regions with a significant forested landbase and this trend is projected continue into the future. US forests cover over 300 Mha, Canadian forests cover over 310 Mha and Russian forests cover approximately 780 Mha (see chapter: Quantifying Forest Biomass Mobilisation Potential in the Boreal and Temperate Biomes). All three countries have seen a decline in markets for conventional forest products (ie timber, pulp and paper, engineered wood products) over the 2000–2010 decade. Demand from the energy sector for woody biomass is seen as a means of diversifying the forest product portfolios in these countries.

As previously mentioned, wood pellet production and trade in the industrial sector largely occurs directly between buyer and seller, with very limited brokerage activity. Production centres across North America and Russia are in direct price competition with each other. Production prices are influenced by:

- The availability of low cost woody biomass and/or residues from existing forestry, pulp and paper, or wood processing industries; and
- The export capacities of the forest or wood processing industries, including infrastructure for bulk shipments (railways, deep-sea harbours) and handling equipment (chippers, cranes, terminals etc.).

Table 8.5 illustrates expected free-on-board prices by 2020, across a selection of harbours, indicating the price advantage of harbours in the South-eastern United States and on the Baltic Sea, compared with Canadian inland shipments from Ontario.

As several independent projections show (Fig. 8.5), the United States (especially the Southeast) is poised to become the largest global producer of wood pellets by 2020. Projected estimates of wood pellet production in the United States vary from 5 to 11.5 million tonnes, depending on the assumed success rate of currently proposed expansion in wood pellet production capacity. By 2020, annual wood pellet production for export is expected to reach between 2.6 and 4.6 million tonnes in Canada and between 1.8 and 7 million tonnes in Russia.

TABLE 8.5 Expected Free-on-Board (FOB) Prices for Wood Pellets Across Selected Harbours in North America and Europe by 2020 (Lamers et al., 2014c)

Harbour	Country/region	FOB (€ tonne/WP$_e$)	FOB (€/GJ)
Halifax (Nova Scotia), Campbellton (New Brunswick)	Eastern Canada (coast)	117	6.65
Montreal (Quebec), Quebec City (Quebec)	Eastern Canada (inland)	131	7.44
Vancouver, Prince Rupert (British Columbia)	Western Canada (coast)	105	5.97
Sankt-Petersburg, Vyborg	North-western Russia (Baltic Sea)	123	6.99
Mykolaiv	Ukraine (Black Sea)	123	6.99
Portland (Maine)	North-eastern USA	117	6.65
Norfolk (Virginia)	Eastern USA	117	6.65
Savannah (Georgia), Mobile (Alabama)	South-eastern USA	108	6.14

Prices per GJ are based on a heating value of 17.6 GJ$_{LHV}$/tonne.

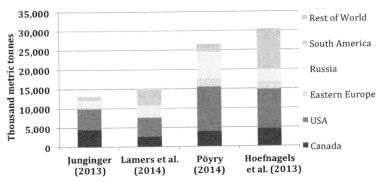

FIGURE 8.5 **Expected annual wood pellet production for international trade by 2020** (Hoefnagels et al., 2013; Junginger, 2013; Lamers et al., 2014c, 2015; Pöyry, 2014). Note: The Pöyry (2014) projection is based on the assumption that 50% of the expected additional capacity by 2025 is already online by 2020. All additional production capacity by Pöyry (2014) is expected to be used for international supply/trade.

SUSTAINABILITY REQUIREMENTS AND CERTIFICATION STATUS

Sustainability and Certification Requirements in Key Industrial Demand Markets in the EU

In 2014, the European Commission announced that it will not pursue binding sustainability criteria to regulate the production of woody biomass for energy purposes before 2020 (EC, 2014). However, numerous regulatory initiatives already exist at the member state and company level. With the aim of importing wood pellets from overseas, large power and heat production utilities from across Northwest Europe (including Electrabel, Dong, Nuon, RWE/Essent/npower, Vattenfall, E.ON) initially formed the International Wood Pellet Buyers Initiative (IWPB), which has been institutionalised as the Sustainable Biomass Partnership (SBP). The SBP sets forth minimum quality requirements for wood pellets (Ryckmans, 2013), which are reflected in the voluntary schemes used by several large utilities, including Electrabel in Belgium (Laborelec scheme) and Essent/RWE/npower in the Netherlands (Green Gold Label).

Despite these initiatives, the European Commission's proposition (EC, 2010) that individual member states adopt solid biomass requirements similar to those for liquid biofuels in RED 2009/28/EC remains valid. Indeed, RED, or similar, criteria are likely to be adopted by individual nations across Northwest Europe and, in case of the United Kingdom, adoption of these criteria has already been proposed (DECC, 2013a; OFGEM, 2011). The Dutch energy industry and nongovernmental organisations (NGOs) have been working towards the development of a national energy accord since the beginning of 2014, and achieved principal agreement on sustainability criteria for solid biomass in March 2015. However, further discussions are required to define a timeline for compliance and to outline how compliance will be tested and monitored (ie which existing SFM certification systems will be used to assess compliance). In the Flemish region of Belgium, a proposal has been prepared to bring the sustainability requirements for woody biomass to the same level as bioliquids (Pelkmans, L., 2014. RE: IEA Bioenergy Task 40 Representative of Belgium, Personal Communication). In Denmark, an Industry Agreement was established in December 2014 to ensure that the use of wood pellets and wood chips for energy production is compliant with the same framework for sustainability (which addresses the environment, health, safety and climate) as CHP producers (Dansk-Energi, 2014).

Critics have claimed that the RED criteria are not adequate for forest biomass, primarily because the suggested GHG emission accounting rules (EC, 2010) would neglect a temporal imbalance between carbon sequestration and release from forest biomass (Searchinger, 2010; Zanchi et al., 2010). This concept, typically referred to as 'carbon debt', may be adopted by aforementioned initiatives that are already in place at member states or company level to augment RED requirements. While it is not entirely clear how such a criterion could impact biomass supply, a rather drastic proposition could be to exclude roundwood, including low-grade pulpwood, from bioenergy production via a feedstock blacklist (see also chapter: Environmental Sustainability Aspects of Forest Biomass Mobilisation). In this case, neither local nor imported wood pellets produced from such feedstocks would be eligible for use.

In an extreme case, the discussion around carbon debt may lead to the exclusion of all forest biomass as a source of energy in large-scale, non-residential applications. This could be achieved by imposing a temporal carbon criterion. In an effort to safeguard biodiversity, the exclusion of solid biomass from 'primary forests', as defined in the RED, could also preclude the use of wood from some Canadian and Russian forests for bioenergy because forestry operations are still sometimes conducted in stands that are 'inherited from nature' and have never been previously harvested on an industrial scale.

The United Kingdom

UK policy support schemes for which sustainability criteria development is relevant include the Renewables Obligation (RO), which is the main support mechanism for large-scale renewable electricity projects; the Renewable Heat Incentive (RHI), which includes domestic incentives for homeowners, private landlords, social landlords, self-builders and non-domestic incentives for industry, businesses and public sector organisations; the Contracts for Difference (CFD), which are long-term contracts to encourage investment in new, low-carbon energy generation; and the Renewable Transport Fuel Obligation, which supports the UK government's policy on reducing GHG emissions from vehicles by encouraging the production of biofuels.

The UK Bioenergy Strategy of April 2012 (DECC, 2012) is based on four principles:

- Policies that support bioenergy should deliver genuine carbon reductions that help meet UK carbon emissions objectives up to 2050 and beyond (a domestic GHG emissions reduction of at least 80% by 2050, against a 1990 baseline).
- Support for bioenergy should make a cost-effective contribution to UK carbon emissions objectives in the context of overall energy goals.
- Support for bioenergy should maximise the overall benefits and minimise costs (quantifiable and non-quantifiable) across the economy.
- At regular time intervals and when policies promote significant additional demand for bioenergy in the UK, policy makers should assess and respond to the impacts of this increased bioenergy consumption on other areas, such as food security and biodiversity.

Since 2012, the UK government has held a number of consultations to develop sustainability criteria for solid biomass and aims to bring in sustainability criteria for those supplying biomass (wood fuel) under the RO and RHI. These criteria were initially introduced as reporting requirements, beginning in April 2014, and as mandatory criteria, beginning in April 2015, for generators above $1MW_{el}$ capacity (OFGEM, 2014). These criteria will apply from the beginning of the CFDs and include:

- A minimum 60% GHG emissions reduction against the average EU fossil grid intensity by 2017, applying the methodology suggested in EC (2010), increasing to 75% by 2025; the methodology considers the emissions from the cultivation, harvesting, processing and transport of the biomass feedstocks. It also includes direct land-use change where the land use has changed category since 2008. It does not include indirect impacts such as displacement effects.
- Land criteria for raw wood and other non-waste biomass, as well as a requirement to source wood from sustainably managed forests, in line with the UK Timber Standard (DECC, 2014) and regardless of where the timber originated.

Since April 2015, all raw wood or biomass products made from raw wood have been managed according to SFM criteria that correspond to the land management criteria for these feedstocks. The SFM criteria are based on the UK Timber Procurement Policy, originally developed to define legal and sustainable timber procurement policies for governmental offices; these are being implemented by the Central Point of Expertise on Timber (CPET). The UK Timber Procurement Policy requires one of two types of evidence to demonstrate that at least 70% of all timber (or biomass) is legally and sustainably harvested:

- *Category A evidence*: certification either through the Forest Stewardship Council (FSC) or an accredited scheme under the Programme for the Endorsement of Forest Certification (PEFC), which currently includes the

Sustainable Forestry Initiative (SFI), the Canadian Standards Association (CSA) and the American Tree Farm System (ATFS).
- *Category B evidence*: bespoke evidence, including the use of a risk-based regional approach, to demonstrate compliance, which covers chain of custody from the forest source to the end-user and which relies on evidence gathered from forest management plans, applicable legislation, supplier declarations, second-party supplier audits and third-party verification.

The UK criteria allow for mixing of feedstocks with different sustainability characteristics at any step in the supply chain, for both Category A and Category B materials, using a mass balancing option.

It is important to note that the current set of UK criteria does not directly address the preservation of land/forest carbon stocks, except where biomass procurement would be classified as a direct land-use change. However, the issues of sustained land/forest carbon loss (carbon debt) and indirect land-use change (ILUC) are being investigated and respective criteria may be integrated in 2016/2017, under the UK Bioenergy Strategy Review (DECC, 2013a).

In a recent comparison of existing SFM certification schemes with the UK Category A and B evidence requirements, Sikkema et al. (2014a) showed that many schemes already comply with most (but not all) requirements (Table 8.6). The GHG emission savings calculations, required by the UK Category A and B evidence requirements, are not adequately addressed by many pre-existing SFM certification schemes. However, this information could be supplied via the OFGEM GHG calculation tool (https://www.ofgem.gov.uk/publications-and-updates/uk-solid-and-gaseous-biomass-carbon-calculator) or a suitable alternative.

Belgium

Currently, there is neither federal support nor binding sustainability requirements for the use of solid biomass to produce heat in Belgium. There is, however, federal support (in the form of support certificates[1]) for ensuring the sustainability of biomass used to produce electricity, including heat from combined heat and power stations. The principles and criteria used to ensure the sustainability of all renewable energy sources (including solid biomass) differ somewhat among the regional quota systems (green certificates) in Flanders, Wallonia and Brussels-Capital. Nevertheless, they all directly link to renewable energy use and GHG reductions based on quota obligations, with special conditions for each region. For example 8 and 15 TWh of renewable energy

---

1. In Belgium, there are two forms of support certificates. In Walloon and Brussels-capital regions, one support (green) certificate is issued for every MWh divided by the amount of CO_2 saved and the certificates are allocated by the regulatory authorities CWaPE and Brugel. In Flemish region, the amount of electricity to be produced for one certificate varies across technologies and is based on a technology-specific banding factor. In general, banding factor is 1 for an amount of 1 MWh of solid biomass use, but banding factor is set at 0.00496 for an amount of 20.161 MWh of biomass used in households and commercial units.

TABLE 8.6 Benchmarking SFM Certification Schemes Against Category A and B Evidence Requirements (Sikkema et al., 2014a)

| | Certified biomass via programs for certified forest management areas (UK Evidence A) | | | | | Miscellaneous options | Complementary programs (UK Evidence B) | | |
| | FSC forest management | PEFC endorsed forest management frameworks | | | | WWF gold standard | FSC CW controlled wood | PEFC due diligence | SFI fibre sourcing |
		PEFC international forest management	SFI forest management	CSA	ATFS	Complementary to FSC			
I. Legal sourcing; EU Timber Regulation (EUTR) (European Commission, 2010a)									
A. Basic compliance: prevention of illegal harvesting practices	✓	✓	✓	✓	✓	✓	✓	✓	✗
II. Sustainable sourcing: EU Communications (European Commission, 2010b)									
A. GHG for forest operations, anticipating a GHG savings requirement	✗	✗	±	±	✗	±	✗	✗	±

(Continued)

TABLE 8.6 Benchmarking SFM Certification Schemes Against Category A and B Evidence Requirements (Sikkema et al., 2014a) *(cont.)*

	Certified biomass via programs for certified forest management areas (UK Evidence A)						Complementary programs (UK Evidence B)		
			PEFC endorsed forest management frameworks			Miscellaneous options			
						WWF gold standard			
	FSC forest management	PEFC international forest management	SFI forest management	CSA	ATFS	Complementary to FSC	FSC CW controlled wood	PEFC due diligence	SFI fibre sourcing
B. No harvest from high biodiversity areas, including primary forest	✓	✓	✓	±	✓	✓	✓	✓	✗
C. No harvest from high carbon stocks or from wetlands	±	±	±	✗	✗	✓	✗	✗	✗
D. Sustainable harvest rates and carbon stocks	✓	✓	✓	±	±	✓	✗	✗	±

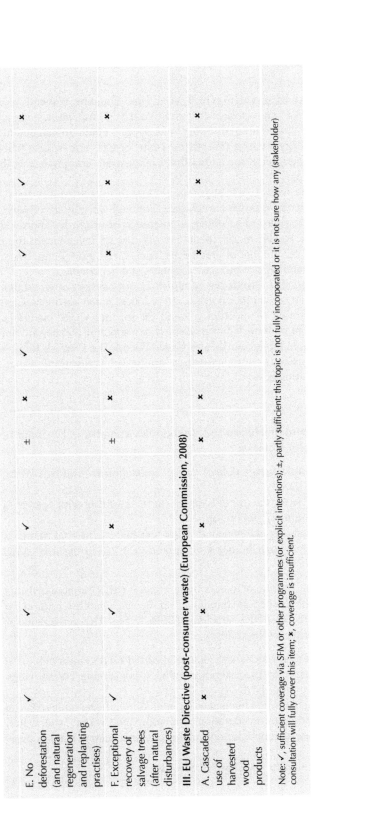

| E. No deforestation (and natural regeneration and replanting practises) | ✓ | ✓ | ✓ | ✗ | ✓ | ✓ | ✓ | ± | ✗ |
| F. Exceptional recovery of salvage trees (after natural disturbances) | ✓ | ✗ | ✓ | ✗ | ✓ | ✗ | ✗ | ± | ✗ |

III. EU Waste Directive (post-consumer waste) (European Commission, 2008)

| A. Cascaded use of harvested wood products | ✗ | ✗ | ✗ | ✗ | ✗ | ✗ | ✗ | ✗ | ✗ |

Note: ✓, sufficient coverage via SFM or other programmes (or explicit intentions); ±, partly sufficient: this topic is not fully incorporated or it is not sure how any (stakeholder) consultation will fully cover this item; ✗, coverage is insufficient.

must be used for electricity production in Wallonia and Flanders, respectively, to meet the target of 13% renewable energy use by 2020. These quota systems include principles of sustainable sourcing and supply chain management, measured via total energy balance (Flemish system) or CO_2 emissions (Walloon system) and requirements for audits that demonstrate compliance with sustainability principles.

There are no requirements to use a specific (voluntary) sustainability standard or certification scheme to prove compliance, although several are eligible. Producers of green electricity must obtain a guarantee of origin by approved certification bodies. However, requirements are verified by independent third parties who determine the amount of green certificates on a case-by-case basis, depending on the size and type of biomass suppliers and generators.

The Flanders region also forbids the combustion (for energy) of wood that could be able to be used for another purpose. In practice, green certificates are awarded for the generation of electricity based on specific waste materials. Waste materials that can be recycled or processed in a superior manner are not accepted for certification. Green certificates issued outside the Flemish Region are not accepted. Discussions on implementing similar wood cascading principles are ongoing in Wallonia.

The Netherlands

The Dutch Energy Accord (published in 2013) outlines a number of key requirements for the use of solid biomass:

- The share of renewable energy in final energy consumption must be 14% by 2020 and 16% by 2023.
- Sustainability criteria for the use of biomass during co-firing with coal are a prerequisite for continuing policy support.
- Co-firing of biomass, including wood chips and pellets, in coal plants is capped at 25 PJ, thus contributing a maximum of 1.2% to the total 2020 target of 14%.

In March 2015, the Dutch socio-economic council (SER) announced that industry and NGOs had reached an agreement on the sustainability criteria for biomass required to receive SDE+ subsidies (SER, 2015). The agreement includes the following criteria (NEA, 2015):

- Criteria for climate and bioenergy: reduction of net GHG emissions (eg a 70% reduction relative to EU reference values), conserving carbon stock reservoirs and preventing ILUC.
- Criteria for sustainable forest management, including criteria on legislation and regulation; ecological considerations (including biodiversity, soil, water, ecological cycles etc.); economic considerations and management considerations.
- Criteria on how to monitor the chain of custody.

● An assessment table (ie a positive/negative list) to include (low carbon debt risk) or exclude (high carbon debt risk) specific materials as bioenergy feedstock. This assessment table is unique within the EU.

Other elements of the agreement include the requirement that large forest management units (ie more than 500 ha) must meet these criteria as of 2015, while for small forest management units, for a limited time (which is not specified), for a number of criteria, a risk-based approach may be applied and only the pellet mill producing wood pellets from these small units needs to be certified. In addition, utilities must create a fund (using the revenues from co-firing wood pellets) to increase SFM certification in the sourcing areas. At the time of writing (Aug. 2015), there was no clarity on how compliance will be tested and monitored, for example which existing SFM certification system will be approved as proof of meeting which criteria, but work is underway to address these gaps.

Denmark

The Danish Energy Agreement (DMCEB, 2012) has set a target of 100% renewable energy use by 2050. To meet this target, a significant increase in the use of renewable energy is required in all sectors, but solid biomass is likely to account for over 50% of the expected total renewable energy consumption.

A voluntary sustainability assurance framework for solid biomass was established in Dec. 2014, based on an initiative by the Danish Energy Ministry, the Danish District Heating Association and the Danish Energy Association. These voluntary sustainability requirements and criteria were developed using the UK Timber Standard for Heat and Electricity (DECC, 2014) as a baseline and reflect the content of the Danish Ministry of the Environment's guidelines on securing sustainable timber in public procurements of goods and services and Forest Europe's criteria for sustainable forest management. The agreement aims to ensure 40 and 100% sustainable biomass use for bioenergy by 2016 and 2019, respectively. The agreement is supplemented by criteria guaranteeing minimum CO_2 savings, compared to fossil alternatives. Participating organisations are also required to evaluate the agreement in 2018 and ensure that it remains consistent with other sustainability frameworks, for example as common sustainability requirements are adopted within the EU or across the globe.

With regard to the production and purchase of wood pellets and wood chips the certification system developed by the SBP can be used, but other voluntary forest certification schemes, such as the FSC or PEFC, are also recognised by the Danish Nature Agency and are considered valid.

The agreement applies to all energy plants that generate heat and electricity using biomass. To ensure that the agreement does not incur disproportionately high costs for smaller facilities, only plants with an input rating exceeding 20 MW_{th} will be subject to documentation requirements, which

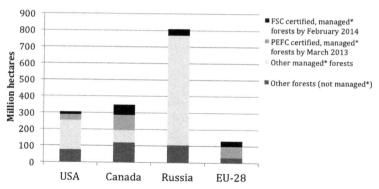

FIGURE 8.6   Forest area covered by certification in North America, Russia, and Europe (NRCAN, 2014; Sikkema et al., 2014a). Note: * Managed forests are defined as total forest minus protection forests, conservation forests and forests preserved for other social services.

will enter into force by August 2016 (Dansk-Energi, 2014). The agreement will be fully phased-in through 2019 and energy utilities which voluntarily participate in this agreement are required to submit annual compliance reports.

Supply-side Certification Status and Volume Estimates

Only a small fraction of the total global forest land mass has been certified (Rekacewicz et al., 2009). However, it is important to note that not all forested land is actively managed. Furthermore, the fraction of certified land varies greatly among countries (Fig. 8.6). The highest certification volumes are based in boreal and temperate forests in Northern Europe, Canada and the United States. By 2012, 151 million ha (Mha) of forest were FSC certified; 88% of this was temperate and boreal forest in North America, Europe and Russia (FSC, 2012). FSC certification covers 6% of the managed forest area in the United States, 18% of that in Canada, 5% of that in Russia and 32% of that in the EU-28. Russia aside, certification under the PEFC umbrella scheme (eg SFI in the United States and CSA in Canada) is much higher. PEFC certification covers 15% of all managed forest area in the United States, 39% of that in Canada and 68% of that in the EU-28 (Fig. 8.6).

Canada

Canadian forests are located predominantly on Crown land, owned and controlled by the government. About 66% of the total Canadian forest landbase of 348 Mha is classified as 'managed forest', that is forest that is under some form of management (NRCAN, 2014). However, the operational definition of 'managed forest' varies from province to province and can include forests in which harvesting has occurred; forests that have never yet been harvested but will be in the future; and forests that will not be harvested but will be protected from wildfires etc. Wood volumes or forested areas are typically allocated to private companies or other entities through contracts. Unlike most European forests, some 'managed' Canadian forests have not yet been harvested and/or are inaccessible by road. There are also vast stretches of forest that are classified as unmanaged although other industrial activities, such as mining, might occur. Most of the managed forest in Canada has been certified by at least one SFM scheme.

Depending on the manner in which the definition of 'primary forest' in the RED is interpreted and applied, biomass harvested from Canadian forests that have not previously been harvested or accessed by roads may be ineligible for use in energy production because they are considered 'natural'. At the same time, Canadian regulations and third-party certification are in place to ensure that ecosystems with high biodiversity value are protected during forest management activities and that forests are managed to conserve characteristics of natural forests, at both the stand and landscape levels, for example by using forest management models that attempt to emulate natural disturbance patterns (Thiffault et al., 2015). Furthermore, a restriction on the use of wood from 'primary forests' for bioenergy is scientifically controversial (Lamers et al., 2013) because this material can still be used for other purposes, such as pulp and paper.

The controversy associated with the perception and definition of 'primary forest' highlights a wider issue, namely fundamental differences in forest management history and practices in Europe and North America. Forest management in Canada is generally conducted in forests inherited by nature and/or influenced by large-scale natural disturbances, such as wildfires. Canadian forestry practices and definitions reflect this reality and contrast strongly with those in Central Europe, where nearly all forests have been managed in an 'unnatural' condition for many decades. This conceptual disconnect could lead to differences in sustainability criteria on the two sides of the Atlantic which could, in turn, result in the imposition of trade barriers on forest biomass.

Most Canadian wood pellets are exported from the province of British Columbia, where feedstocks used to produce wood pellets include sawdust and shavings from timber processing, harvesting residues and salvaged wood. Salvaged wood comes predominantly from trees killed by a catastrophic outbreak of mountain pine beetle (*Dendroctonus ponderosae*), which have since become unmerchantable as timber (BC-MoF, 2010a, 2010b). Technically, this wood can be classified as 'roundwood', that is stemwood, but forest companies sell merchantable timber to the conventional wood products industry and unmerchantable trees (based on stem size, shape, or the presence of fungal stains) to wood pellet producers.

In the eastern Canadian provinces of Ontario and Quebec, increasing volumes of roundwood are used in the wood pellet industry due to the closure of pulp and paper mills. Stakeholders in these provinces are also considering the use of low-quality trees salvaged after natural disturbances (mainly wildfire and insects, for example spruce budworm, *Choristoneura fumiferana*) for pellet production (Barrette et al., 2015).

Numerous estimates of forest biomass availability have been calculated for Canadian forests (Paré et al., 2011), focusing primarily on residues, trees and tree parts that are not used by conventional forest industries. To date, these estimates have not included unutilised annual allowable cut, although these trees could also be considered as a potential source of bioenergy feedstock. A comprehensive analysis of biomass feedstocks for energy production in Canada was performed by Dymond et al. (2010), who considered the bioenergy potential of residues from clear-cutting (slash), forest fires, insect outbreak, stand break-up and self-thinning. Dymond et al. (2010) estimated that approximately 52 million tonnes of biomass is available each year from clear-cut residues, wildfires and insect disturbances. An additional 178 million tonnes per year may be available from stand break-up and self-thinning. Although a detailed sustainability assessment was not included, the authors did consider a 50% outtake ratio, that is 50% of all available biomass would remain in the forest (Dymond et al., 2010).

The United States

Certification in the United States is not as widespread as it is in Canada, due, in large part, to diverse patterns of forest ownership. Large areas with private, small-holder ownership are common in the Southeast. By contrast, large, continuous stretches of corporate-owned forests exist in the Northeast (eg Maine), the Northwest (eg Oregon and Washington) and in Texas and Louisiana (USFS, 2011).

SE United States currently supplies 60% of the total volume of wood harvested in the country and is expected to remain the key wood-producing region for the foreseeable future (DOE, 2011). Two-thirds of these forests are owned by non-industrial private forest landowners (Pinchot-Institute, 2013). Timberland owned and controlled by the forest industry is less common in this region. At a recent forest and trade workshop between industry, government and NGOs, participants noted that only 3% of non-industrial private forest landowners in the Southeast have a written forest management plan and only 13% have received forest management advice (Pinchot-Institute, 2013). To many of these landowners, income via timber harvesting is only one of several ownership objectives, which also include providing wildlife habitat and hunting revenues. Small-holder, non-industrial private forest land owners are reluctant to pay for voluntary forest certification, which is reflected in the relatively low overall SFM certification level of 17% (Pinchot-Institute, 2013).

SE United States is the current and expected primary sourcing region for wood pellet exports to the key demand regions as identified in the previous section (Fig. 8.7). By 2020, over 90% of the US wood pellet production capacity for export is expected to be in this region. While there has been reported use of hardwoods (http://wunc.org/post/advocates-report-critical-nc-wood-pellet-mill), the

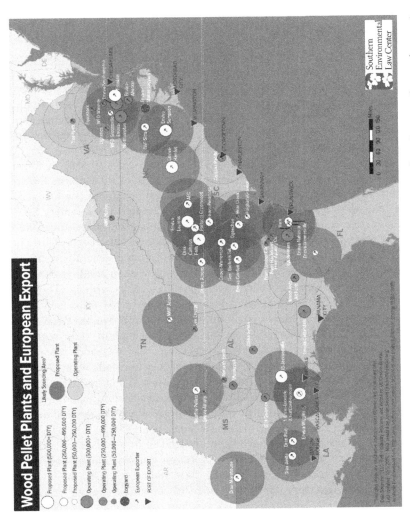

FIGURE 8.7 Current and expected US wood pellet production for export (http://www.southernenvironment. org/cases-and-projects/biomass-energy-in-the-south).

vast majority of the feedstock will be small diameter and pulpwood quality roundwood from pine plantations. Additional feedstock options include timber harvest residues, that is primary residues such as tops and branches (used, eg by Georgia Biomass) and, to a limited extent, also processing residues such as sawdust and shavings (used, eg by Fram Renewables). For an extended list of wood pellet plants and their capacities, see Hess et al. (2015).

Russia

One of the world's largest wood pellet plants is located in Vyborg, Northwest Russia, the country's main production region of wood pellets destined for export to Europe. In fact, between 30 and 50% of the total annual Russian timber production originates in Northwest Russia. This region has long-standing relationships with European export markets because of its proximity to the border and to shipping routes on the Baltic Sea (Thiffault et al., 2014). Wood pellet production for markets in Asia has also recently emerged in eastern Russia (Cocchi et al., 2011).

According to the Russian Federal Agency of Forestry, 120 Mha of forest are SFM certified; this represents approximately 25% of all Russian forests (Thiffault et al., 2014). By 2014, the majority of forest operations conducted by foreign companies across Northwest Russia were certified via voluntary SFM labels. However, management and harvesting practices within these certified forests are often not in compliance with respective SFM requirements (Thiffault et al., 2014), suggesting that third party SFM certification is generally weak. In some cases, the application of FSC standards has resulted in conflicts between Russian forest legislation and management practices on the ground (Thiffault et al., 2014). However, this apparent lack of effective SFM implementation may be improving. Since March 2013, for example the EU Timber Regulation (http://ec.europa.eu/environment/forests/timber_regulation.htm) has required traders in wood pellets to exercise due diligence (eg chain of custody reporting) for any wood imported into the EU. Wood for energy also falls within the application of the timber regulation, except when classified and traded under the respective trade code of 'waste wood'. Past uses of trade codes (including those for waste wood) have shown that definitions will eventually determine the effectiveness of respective trade regulations (Lamers et al., 2012).

Few peer-reviewed studies have been undertaken to estimate the volume of wood available for energy production in Russia, particularly in the context of sustainability restrictions. Based on the 2004 annual cut, Gerasimov et al. (2007) calculated that approximately 4 Mm^3 of thinnings, logging residues, non-industrial roundwood and secondary residues would be available for bioenergy in the St Petersburg region. The majority of this material (approximately 86%) was expected to come from non-industrial roundwood and felling residues, with the remainder from secondary/mill residues. Based on the results of this study, it may be possible to double the utilisation of forest resources for energy production, if the annual allowable cut (AAC) was fully harvested and thinnings utilised (Gerasimov et al., 2007). Goltsev et al. (2010) updated these

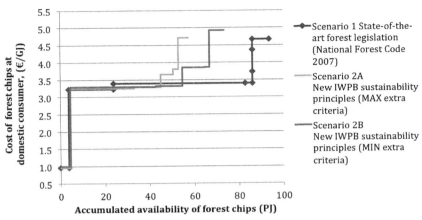

FIGURE 8.8   Cost supply curves for forest chips in the Leningrad region in 2006 (Sikkema et al., 2014b).

scenarios for several districts in the St Petersburg region and found that almost half the potential wood volume is cut annually and that the volume of biomass for energy could increase by 83%, should the potential cut volume be fully utilised. However, sustainability was not considered in their analysis.

Sikkema et al. (2011) estimated that 95 PJ (equivalent to 13 Mm^3, assuming 7.3 GJ/m^3 of solid content of wood chips) of energy could be produced from forest resources in the Leningrad region, if utilisation was limited by the criteria set of the SBP. Potential production was reduced to 54–58 PJ if additional criteria, such as limits on the use of roundwood, stumps and harvesting residues were applied, or if additional protected forests were set aside (Scenario 2A, Fig. 8.8). If roundwood was used for energy and ash was applied after stump and slash removal, however, potential production was only reduced to 74 PJ (Scenario 2B, Fig. 8.8).

DISCUSSION: SUPPLY LIMITATIONS AND POTENTIAL TRADE BARRIERS

In addition to the barriers to mobilisation observed in supply chains at the local and regional scale and discussed in previous chapters, at present, limitations on the supply of woody biomass available for international trade from Canada, the United States and Russia are related to potential trade restrictions, in the form of regulatory measures. The selection and definition of sustainability criteria will have a strong influence on regulatory trade restrictions.

Whether sustainability criteria, such as those outlined in the United Kingdom's Timber Procurement Policy, infringe on international agreements under the World Trade Organization (WTO), such as the Technical Barriers to Trade Agreement, has been discussed in the literature (Burrell et al., 2012; Mitchell and Tran, 2009). For the WTO, the concept of sustainability hinges on whether

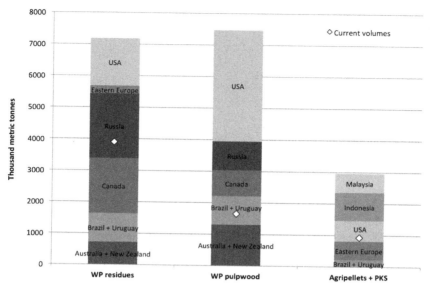

FIGURE 8.9   **Global wood pellet** *(WP)*, **agripellet and palm kernel shell** *(PKS)* **production for energy export markets by 2020 (Lamers et al., 2015).**

traded goods can be distinguished by how they are made [process and production methods (PPM)] or merely by their physical attributes (Goetzl, 2015). The applicability of WTO rules to products differentiated by PPM remains an issue of active debate that is largely unresolved (Goetzl, 2015; Mitchell and Tran, 2009).

The supply of woody biomass from Canada could be strongly affected by the imposition of specific sustainability criteria and definitions such as the description of 'primary forests' put forth in the RED 2009/28/EC (Thiffault et al., 2014). A prohibition on the use of roundwood could also reduce the quantity of Canadian wood pellets produced for export. In British Columbia, where a fraction of the wood pellets produced for export are currently from processing residues of mountain pine beetle-killed stands, but also from dead trees that are no longer of sawlog quality, this limitation could reduce the output of wood pellets by at least 1 million tonnes per year, by 2020 (Fig. 8.9).

In the United States, the minimum sustainability requirements put forward by the UK's Timber Procurement Policy may lead to an increase in SFM certification, or to the collection of similar evidence provided via Category B designations (eg forest management plans, applicable legislation, supplier declarations, second-party supplier audits, third-party verification). A feedstock blacklist that excludes the use of roundwood, including low quality pulpwood, would reduce the volume of absolute projected wood pellet production for export in the SE United States, from an estimated 5–6 million tonnes by 2020, down to 1–2 million tonnes (Fig. 8.9).

A recent analysis investigating the effect of international policies on the industry and forests in SE United States indicated that the key drivers for the continuous growth of local wood pellet production include the characteristics of the current forest inventory and EU definitions of sustainability (Abt et al., 2014). Requirements for minimum GHG emission reductions, land-use change avoidance and SFM certification will reduce the available inventory and increase feedstock prices (Abt et al., 2014). When the demand for wood pellet feedstock increases, the price-inelastic demand and supply response is expected to result in price hikes. The precise impact of feedstock shortages will depend on the capability of respective buyers to pay increased costs, on policy support schemes in Europe and on the demand in local US markets for timber.

In Russia, the imposition of sustainability criteria is expected to be a mix of the situations in Canada and the United States. An exclusion of 'primary forest' would limit the forest landbase from which feedstock could be gathered. At the same time, the UK requirements for SFM may encourage certification. Given that aspen and other residual trees that are typically not collected for use by the pulp and paper or timber industries will qualify as forest residues, Russia may still be able to generate 3 million tonnes of wood pellets for export, by 2020. However, a restriction on the use of pulpwood-quality roundwood would reduce the output volume at least by 1 million tonnes (Fig. 8.9).

Canada, the United States and Russia have strongly contrasting governance structures and forestry contexts. An analysis by Thiffault et al. (2014) of the potential impact of anticipated European sustainability criteria on the national and local regulations in these countries illustrates potential challenges including:

- Differences between jurisdictions in land definitions, delineation and reporting systems;
- a lack of a uniform paradigm for SFM; and
- difficulties in establishing efficient monitoring/auditing systems.

Consensus is emerging on the need to account for biogenic carbon emissions over time. However, the principles and outcomes considered necessary to do so vary considerably among stakeholders. While there appears to be agreement that the carbon emitted through the combustion of biomass for energy was and will again be sequestered from the atmosphere (if the quantity of biomass used can be associated with the regrowth of a forest in a SFM system), there is concern about the time lag between carbon release and carbon (re-) sequestration (see the chapter: Environmental Sustainability Aspects of Forest Biomass Mobilisation). This temporal carbon imbalance is particularly relevant for forest ecosystems that require longer rotation cycles, such as boreal forests.

The sustainable use of woody biomass for energy requires that GHG accounting considers alternative biomass and land uses and compares these with the systems using fossil fuels. This is particularly true for the combustion of roundwood. Such accounting exercises are less critical to justify the utilisation of residues or co-products from timber harvesting and/or wood processing industries, which operate independently from energy markets and incentives. A large fraction of these residues would either decay naturally or be burned, landfilled etc.

if they were not used for the production of energy (see the chapter: Quantifying Forest Biomass Mobilisation Potential in the Boreal and Temperate Biomes). Therefore, much of the biogenic carbon contained in these residues would be released back into the atmosphere relatively quickly, compared to carbon stored in wooden construction, for example. It will be critical to define these residues correctly because their increased utilisation for biomass may lead to unsustainable rates of harvest, or the use of less efficient processing technologies and greater waste.

Policy options to address biogenic carbon emissions include mechanisms that quantify associated emissions, such as the integration of forest carbon accounting in full life-cycle assessments. Preventative approaches include requirements for SFM that guarantee replanting and sustained carbon stocks/yields and actively discouraging the conversion of specific lands (eg peatlands, where drainage releases large amounts of GHG). Some stakeholders have also suggested feedstock/biomass blacklists and/or cascading policies. It should be noted, however, that the exclusion of certain feedstock fractions or non-SFM certified material does not prevent its leakage into other markets (eg pulp and paper) or regions with less stringent sustainability criteria (eg Asia).

SUMMARY AND CONCLUSIONS

The global annual demand for internationally traded wood pellets from boreal and temperate forests for heat and power generation is expected to reach 15–26 million tonnes (264–458 PJ) by 2020. The European Union, particularly the United Kingdom, the Netherlands, Denmark and Belgium will remain the key destination markets. Although demand from Asia (South Korea and Japan) will grow, this region will play a secondary role for the foreseeable future.

Adoption of RED, or similar, criteria is likely across all EU countries and, in the United Kingdom (the largest market for traded wood pellets), such criteria have already been proposed. As of 2015, only forest materials that achieve at least 60% GHG emissions savings against the EU fossil fuel electricity average can be used for bioenergy production and, furthermore, proof of SFM is required. Eligible SFM schemes for the UK market include FSC certification and schemes endorsed by the PEFC, such as the SFI and CSA. In the Netherlands, the energy industry and NGOs have been collaborating on the development of an Energy Accord since 2014 and recently achieved principal agreement on the sustainability criteria for solid biomass. While specific criteria have been laid out, key issues that remain include compliance testing and monitoring. In Flemish Belgium, a proposal is being prepared to bring the sustainability requirements for woody biomass to the same level as bioliquids. In Denmark, a voluntary industry agreement is set to ensure that 40 and 100% of all bioenergy production is conducted sustainably by 2016 and 2019, respectively.

At present, the key supply regions of traded wood pellets from temperate and boreal biomes are (in order of importance) the United States, Canada and

Russia and this is not expected to change by 2020. Among these, Canada offers the largest stretches of SFM-certified forests, but the US Southeast has seen the strongest increase in wood pellet production and export in recent years. Differences between Canada and the United States in the level of SFM certification are largely due to differences in forest ownership (ie public vs. private). The dramatic increase in pellet production in the US Southeast compared with other regions, for example Western or Central Canada, is linked to a number of factors, including available forest inventory (particularly pulpwood fractions) and competitive transport advantages along the eastern coast of North America and thus better access to EU markets. In addition to barriers to mobilisation identified at the local and national scales in previous chapters and related to policy, logistics, conversion technologies etc. supply limitations for international trade will also be influenced by regulatory measures, including sustainability criteria, which will restrict the use of some feedstocks and, thus, trade. In the US Southeast, an increased demand for wood pellets is expected to raise feedstock prices and EU sustainability criteria that limit feedstock options for producers exporting to Europe will further stimulate this trend. The biomass volumes that can eventually be mobilised will thus depend on each country's framework conditions, that is the respective forest biomass sustainability criteria and the underlying policy mechanisms, which define the energy utilities' willingness-to-pay for imported wood pellets. The sustainability definitions with the greatest potential consequences for woody biomass availability are those of 'primary forests' (imposed to protect habitats with a high biodiversity value) and 'residue' or 'co-products' (imposed to ensure the utilisation only of harvesting and processing residues with no alternative use).

REFERENCES

Abt, K.L., Abt, R.C., Galik, C.S., Skog, K.E., 2014. Effect of Policies on Pellet Production and Forests in the US South: A Technical Document Supporting the Forest Service Update of the 2010 RPA Assessment. US Department of Agriculture Forest Service, Southern Research Station, Asheville, NC, Gen. Tech. Rep. SRS-202. Available from: http://www.srs.fs.usda.gov/pubs/gtr/gtr_srs202.pdf.

AEBIOM, 2012. European Bioenergy Outlook—Statistical Report. European Biomass Association, Brussels, Belgium.

AEBIOM, 2013. European Bioenergy Outlook—Statistical Report. European Biomass Association, Brussels, Belgium.

Agentschap-NL, 2013. Green deal Duurzaamheid Vaste Biomassa, Den Hague, Netherlands. Available from: http://www.rvo.nl/onderwerpen/duurzaam-ondernemen/groene-economie/duurzaamheid-vaste-biomassa/publicaties.

Barrette, J., Thiffault, E., Saint-Pierre, F., Wetzel, S., Duchesne, I., Krigstin, S., 2015. Dynamics of dead tree degradation and shelf-life following natural disturbances: can salvaged trees from boreal forests 'fuel' the forestry and bioenergy sectors? Forestry 88 (3), 275–290.

BC-MoF, 2010a. The State of British Columbia's Forests. Ministry of Forests and Range; Forest Analysis and Inventory Branch, Victoria, BC, Available from: http://www.for.gov.bc.ca/hfp/sof/2010/SOF_2010_Web.pdf.

BC-MoF, 2010b. Emergency Bark Beetle Management Area (EBBMA) and Strategic Planning Map: Mountain Pine Beetle. Ministry of Forests, Lands and Natural Resource Operations, Victoria, BC.

Beurskens, L., Hekkenberg, M., 2010. Renewable Energy Projections as Published in the National Renewable Energy Action Plans of the European Member States. European Commission, Petten, the Netherlands, Available from: http://www.ecn.nl/docs/library/report/2010/e10069.pdf.

Blair, L., 2013. Global Pellet Markets—Towards a Commodity Status? Argus Media, Brussels, Belgium, Presented at AEBIOM Conference, Brussels, Belgium, June 2013.

Burrell, A., Gay, S.H., Kavallari, A., 2012. The compatibility of EU biofuel policies with global sustainability and the WTO. World Econ. 35 (6), 784–798.

Chum, H., Faaij, A., Moreira, J., Berndes, G., Dhamija, P., Dong, H., Gabrielle, B., Goss Eng, A., Lucht, W., Mapako, M., Masera Cerutti, O., McIntyre, T., Minowa, T., Pingoud, K., 2011. Bioenergy. In: Edenhofer, O., Pichs-Madruga, R., Sokona, Y., Seyboth, K., Matschoss, P., Kadner, S., Zwickel, T., Eickemeier, P., Hansen, G., Schlömer, S., Stechow, C. v. (Eds.), IPCC Special Report on Renewable Energy Sources and Climate Change Mitigation. Cambridge University Press, Cambridge, UK and New York, NY.

Cocchi, M., Nikolaisen, L., Junginger, M., Goh, C., Heinimö, J., Bradley, D., Hess, R., Jacobson, J., Ovard, L., Thrän, D., Hennig, C., Deutmeyer, M., Schouwenberg, P.-P. and Marchal, D., 2011. Global wood pellet industry and market study: IEA Bioenergy Task 40. Available from: http://www.bioenergytrade.org/downloads/t40-global-wood-pellet-market-study_final.pdf

Creutzig, F., Ravindranath, N.H., Berndes, G., Bolwig, S., Bright, R., Cherubini, F., Chum, H., Corbera, E., Delucchi, M., Faaij, A., Fargione, J., Haberl, H., Heath, G., Lucon, O., Plevin, R., Popp, A., Robledo-Abad, C., Rose, S., Smith, P., Stromman, A., Suh, S., Masera, O., 2014. Bioenergy and climate change mitigation: an assessment. GCB Bioenergy 7 (5), 916–944.

Dale, A. Expanding demand and supply perspectives for the industrial pellet markets. AEBIOM Conference, Brussels, Belgium, June 2013: Ekman.

Dansk-Energi, 2014.Report on Industry agreement to ensure sustainable biomass (wood pellets and wood chips).

DEA, 2012a. Danmarks Energifremskrivning. Danish Energy Agency, Copenhagen, Denmark.

DEA, 2012b. Energy Statistics 2011 Data, Tables, Statistics, and Maps. Danish Energy Agency, Copenhagen, Denmark.

DECC, 2012. UK Bioenergy Strategy. Department of Energy & Climate Change, Department of Transport, Department for Environment, Food and Rural Affairs, London, UK, (URN: 12D/077). Available from: https://www.gov.uk/government/uploads/system/uploads/attachment_data/file/48337/5142-bioenergy-strategy-.pdf.

DECC, 2013a. Government Response to the Consultation on Proposals to Enhance the Sustainability Criteria for the Use of Biomass Feedstocks Under the Renewables Obligation (RO). Department of Energy & Climate Change, London, UK, Available from: https://www.gov.uk/government/uploads/system/uploads/attachment_data/file/231102/RO_Biomass_Sustainability_consultation_-_Government_Response_22_August_2013.pdf.

DECC, 2013b. Energy Trends Section 6 Renewables. Department of Energy & Climate Change, London, UK, Available from: https://www.gov.uk/government/publications/renewables-section-6-energy-trends.

DECC, 2014. UK Timber Standard for Heat and Electricity: Woodfuel Used Under the Renewable Heat Incentive and Renewables Obligation. Department of Energy & Climate Change, London, UK, URN: 14D/025. Available from: https://www.gov.uk/government/uploads/system/uploads/attachment_data/file/278372/Timber_Standard_for_Heat_and_Electricity_under_RO_and_RHI_-_10-Feb-2014_for_pdf_-_FINAL_in_new_format.pdf.

DMCEB, 2012. DK Energy Agreement, March 22. Danish Ministry of Climate, Energy and Building, Copenhagen, Denmark, Available from: http://www.kebmin.dk/sites/kebmin.dk/files/climate-energy-and-building-policy/denmark/energy-agreements/FAKTA UK 1.pdf.

DOE, 2011. US Billion-Ton Update: Biomass Supply for a Bioenergy and Bioproducts Industry. Oak Ridge National Laboratory, Oak Ridge, TN, U.S. Department of Energy (ORNL/TM-2011/224). Available from: http://www1.eere.energy.gov/bioenergy/pdfs/billion_ton_update.pdf.

Dymond, C.C., Titus, B.D., Stinson, G., Kurz, W.A., 2010. Future quantities and spatial distribution of harvesting residue and dead wood from natural disturbances in Canada. Forest Ecol. Manage. 260 (2), 181–192.

EC, 2010. Report on Sustainability Requirements for the Use of Solid and Gaseous Biomass Sources in Electricity, Heating and Cooling. European Commission, Brussels, Belgium.

EC, 2014. Report on State of Play on the Sustainability of Solid and Gaseous Biomass Used for Electricity, Heating and Cooling in the EU. European Commission, Brussels, Belgium, Available from: http://ec.europa.eu/energy/sites/ener/files/2014_biomass_state_of_play_pdf.

European Commission, 2008. Directive 2008/98/EC on waste and repealing certain directives (Waste framework Directive). Off. J. Eur. Union 2008 L312, 3–30.

European Commission, 2010a. Obligations of operators who place timber an timber products on the market (EU Timber Regulation). Off. J. Eur. Union 2010 L295, 23–34, Directive 2010/995/EC.

European Commission, 2010b. Report to the Council and the European Parliament on sustainability requirements for the use of solid and gaseous biomass sources in electricity, heating and cooling. SEC 2010, 1–20, 2010 final.

FSC, 2012. Global FSC Certificates: Type and Distribution. Forest Stewardship Council, Bonn, Germany, Available from: http://ic.fsc.org/download.facts-and-figures-may-2012.a-225.pdf.

Gerasimov, Y., Karjalainen, T., Ilavský, J., Tahvanainen, T., Goltsev, V., 2007. Possibilities for energy wood procurement in north-west Russia: assessment of energy wood resources in the Leningrad region. Scand. J. Forest Res. 22 (6), 559–567.

Goetzl, A., 2015. Development in the Global Trade of Wood Pellets. US International Trade Commission, Washington, DC, (ID-039). Available from: http://www.usitc.gov/publications/332/wood_pellets_id-039_final.pdf.

Goh, C.S., Junginger, M., Cocchi, M., Marchal, D., Thrän, D., Hennig, C., Heinimö, J., Nikolaisen, L., Schouwenberg, P.-P., Bradley, D., Hess, R., Jacobson, J., Ovard, L., Deutmeyer, M., 2013. Wood pellet market and trade: a global perspective. Biofuel. Bioprod. Bioref. 7 (1), 24–42.

Goltsev, V., Ilavský, J., Karjalainen, T., Gerasimov, Y., 2010. Potential of energy wood resources and technologies for their supply in Tihvin and Boksitogorsk Districts of the Leningrad Region. Biomass Bioenergy 34 (10), 1440–1448.

Harrabin, R., Biomass fuel subsidies to be capped says energy secretary. BBC News. Available from: http://www.bbc.co.uk/news/business-23334466

Hawkins-Wright. The Outlook for Wood Pellet Demand. USIPA Third Annual Exporting Pellets Conference, 27–29 October 2013, Miami, FL, USA.

Heinimö, J., 2008. Methodological aspects on international biofuels trade: International streams and trade of solid and liquid biofuels in Finland. Biomass Bioenergy 32 (8), 702–716.

Heinimö, J., Lamers, P. and Ranta, T. International trade of energy biomass—an overview of the past development. Twenty-first European Biomass Conference, Copenhagen, Denmark, June 2013, pp. 2029–2033.

Hess, R., Lamers, P., Roni, M., Jacobson, J., Heath, B., 2015. United States Country Report—IEA Bioenergy Task 40. Idaho National Laboratory, Idaho Falls, ID, Available from: http://www.bioenergytrade.org/downloads/iea-task-40-country-report-2014-us.pdf.

Hoefnagels, R., Cornelissen, T., Junginger, M., Faaij, A., 2013. Capacity Study for Solid Biomass Facilities. Utrecht University, Utrecht, the Netherlands, Available from: http://www.portofrotterdam.com/en/Business/rotterdam-energy-port/Documents/PoRCapacitystudyTradeFlowsFinalApril2013.pdf.

IEA, 2012. World Energy Outlook. International Energy Agency, Paris, France.

IPCC, 2007. Contribution of Working Groups I, II and III to the Fourth Assessment Report of the Intergovernmental Panel on Climate Change. Intergovernmental Panel on Climate Change, Geneva, Switzerland, p. 104.

IPCC, 2014. Climate Change 2014 Mitigation of Climate Change. Contribution of Working Group III to the Fifth Assessment Report of the Intergovernmental Panel on Climate Change. Cambridge University Press, Cambridge, UK and New York, NY.

Junginger, M. Forest carbon accounting on co-firing woody biomass–a matter of perspectives? Joint Workshop on Developing a Binding Sustainability Scheme for Solid Biomassfor Electricity & Heat underthe RED, Arona, Italy, 1–2 July.

Junginger, M., Goh, C.S., Faaij, A. (Eds.), 2013. International Bioenergy Trade: History, Status & Outlook on Securing Sustainable Bioenergy Supply, Demand and Markets. Springer, Berlin, Germany.

Junginger, M., Schouwenberg, P-P, Nikolaisen, L., Andrade, O., 2014. Drivers and barriers for bioenergy trade. In: Junginger, M., Goh, C.S., Faaij, A. (Eds.), International Bioenergy Trade: History, Status & Outlook on Securing Sustainable Bioenergy Supply, Demand and Markets. Springer, Berlin, Germany, pp. 151–172.

Kranzl, L., Daioglou, V., Faaij, A., Junginger, M., Keramidas, K., Matzenberger, J., Tromborg, E., 2014. Medium and long-term perspectives of international bioenergy trade. In: Junginger, M., Goh, C.S., Faaij, A. (Eds.), International Bioenergy Trade: History, Status & Outlook on Securing Sustainable Bioenergy Supply, Demand and Markets. Springer, Berlin, Germany, pp. 173–189.

Lamers, P., 2014. Sustainable international bioenergy trade: evaluating the impact of sustainability criteria and policy on past and future bioenergy supply and trade. PhD Thesis: Energy and Resources, Copernicus Institute, Utrecht University, the Netherlands.

Lamers, P., Junginger, M., Hamelinck, C., Faaij, A., 2012. Developments in international solid biofuel trade—an analysis of volumes, policies, and market factors. Renew. Sustain. Energy Rev. 16 (5), 3176–3199.

Lamers, P., Thiffault, E., Paré, D., Junginger, H.M., 2013. Feedstock specific environmental risk levels related to biomass extraction for energy from boreal and temperate forests. Biomass Bioenergy 55 (8), 212–226.

Lamers, P., Marchal, D., Heinimö, J., Steierer, F., 2014a. Woody biomass trade for energy. In: Junginger, M., Goh, C.S., Faaij, A. (Eds.), International Bioenergy Trade: History, Status & Outlook on Securing Sustainable Bioenergy Supply, Demand and Markets. Springer, Berlin, Germany, pp. 41–64.

Lamers, P., Rosillo-Calle, F., Pelkmans, L., Hamelinck, C., 2014b. Developments in international liquid biofuel trade. In: Junginger, M., Goh, C.S., Faaij, A. (Eds.), International Bioenergy Trade: History, Status & Outlook on Securing Sustainable Bioenergy Supply, Demand and Markets. Springer, Berlin, Germany, pp. 17–40.

Lamers, P., Hoefnagels, R., Junginger, M., Hamelinck, C., Faaij, A., 2014c. Global solid biomass trade for energy by 2020 an assessment of potential import streams and supply costs to North-West Europe under different sustainability constraints. GCB Bioenergy 7 (4), 618–634.

Lamers, P., Hoefnagels, R., Junginger, M., Hamelinck, C., Faaij, A., 2015. Global solid biomass trade for energy by 2020 an assessment of potential import streams and supply costs to North-West Europe under different sustainability constraints. GCB Bioenergy 7 (4), 618–634.

Mitchell, A.D., Tran, C., 2009. The consistency of the EU renewable energy directive with the WTO agreements. Georgetown Law Faculty Working Papers. Paper 119. 15 p. Available from: http://scholarship.law.georgetown.edu/fwps_papers/119

NEA, 2015. SDE+ Sustainability Requirements for Co-firing and Large Scale Heat Production. Netherlands Enterprise Agency, The Hague, the Netherlands, Available from: http://english. rvo.nl/sites/default/files/2015/04/SDE+%2B sustainability requirements for co-firing and large scale heat production.pdf.

NRCAN, 2014. The State of Canada's Forests—Annual Report 2014. Natural Resources Canada, Canadian Forest Service, Ottawa, Canada, Available from: http://cfs.nrcan.gc.ca/pubwarehouse/ pdfs/35713.pdf.

OFGEM, 2011. Renewables Obligation: Sustainability Criteria for Solid and Gaseous Biomass for Generators (Greater Than 50 Kilowatts). UK Office of the Gas and Electricity Markets, London, UK, Available from: http://www.ofgem.gov.uk/Sustainability/Environment/RenewablObl/ FuelledStations/Documents1/SolidandGaseousBiomassGuidanceFINAL.pdf.

OFGEM, 2013. Electricity Capacity Assessment Report 2013. Office of Gas and Electricity Markets, London, UK, Available from: http://www.ofgem.gov.uk/Markets/WhlMkts/monitoring-energy-security/elec-capacity-assessment/Documents1/Electricity Capacity Assessment Report 2013.pdf.

OFGEM, 2014. Renewables Obligation: Sustainability Criteria Guidance. UK Office of the Gas and Electricity Markets, London, UK, Available from: https://www.ofgem.gov.uk/publications-and-updates/renewables-obligation-sustainability-criteria-guidance-0.

Paré, D., Bernier, P., Thiffault, E., Titus, B.D., 2011. The potential of forest biomass as an energy supply for Canada. Forestry Chron. 87 (1), 71–76.

Pinchot-Institute, 2013. The Transatlantic Trade in Wood for Energy: A Dialogue on Sustainability Standards and Greenhouse Gas Emissions. Pinchot Institute for Conservation, Savannah, GA, Available from: http://cif-seek.org/wp-content/uploads/2013/11/Trade-in-Wood-for-Energy_Savannah-Workshop-Summary_Final.pdf.

Pöyry, 2000. The Dynamics of Global Pellet Markets, 4. Central European Biomass Conference. Pöyry Management Consulting, London, UK.

Pöyry, 2014. International Trading—Global Pellet Market Outlook to 2020. Pöyry Management Consulting, London, UK, Presented at Interpellets 2011 Conference, Stuttgart.

Rekacewicz, P., Marin, C., Stienne, A., Frigieri, G., Pravettoni, R., Margueritte, L., Lecoquierre, M., 2009. Vital Forest Graphics. GRID Arendal (a centre collaborating with UNEP), Arendal, Norway, Available from: http://www.grida.no/graphicslib/detail/very-little-forest-area-is-certified_bd83.

REN21, 2014. Renewables Global Status Report. Renewable Policy Network for the 21st Century, Paris, France, Available from: http://www.ren21.net/GSR2014-Renewables-2014-Global-Status-Report-Key-Findings-EN.

REN21, 2015. Renewables Global Status Report. Paris, France, Renewable Policy Network for the 21st Century, Available from: http://www.ren21.net/GSR-2015-Report-Full-report-EN.

Roos, J.A., Brackley, A.M., 2012. The Asian Wood Pellet Markets. US Department of Agriculture, Forest Service, Pacific Northwest Research Station, Portland, OR, Available from: http://www. fs.fed.us/pnw/pubs/pnw_gtr861.pdf.

Ryckmans, Y., 2013. An Introduction to IWPB, Laborelec, Belgium. Available from: http://www. imp.gda.pl/bioenergy/bruksela/laborelec.pdf

Searchinger, T.D., 2010. Biofuels and the need for additional carbon. Environ. Res. Lett. 5 (2), 024007.

SER, 2015. Utilities and NGOs Agree on Sustainability Criteria Biomass [in Dutch]. The Social and Economic Council of the Netherlands, The Hague, the Netherlands, Available from: https:// www.ser.nl/nl/actueel/persberichten/2010-2019/2015/20150313-akkoord-biomassa.aspx.

Sikkema, R., Steiner, M., Junginger, M., Hiegl, W., Hansen, M., Faaij, A., 2011. The European wood pellet markets: current status and prospects for 2020. Biofuel. Bioprod. Bioref. 5 (3), 250–278.

Sikkema, R., Junginger, M., Dam, J., Stegeman, V., Durrant, G., Faaij, D.A., 2014a. Legal and sustainable wood sourcing for bioenergy and cross compliance with certification frameworks for Sustainable Forest Management. Forests 5, 2163–2211.

Sikkema, R., Faaij, A., Ranta, T., Heinimö, J., Gerasimov, Y., Karjalainen, T., Nabuurs, G., 2014b. Mobilization of biomass for energy from boreal forests in Finland & Russia under present sustainable forest management certification and new sustainability requirements for solid biofuels. Biomass Bioenergy 71, 23–36.

Thiffault, E., Lorente, M., Murray, J., Endres, J.M., McCubbins, J.S.N., Fritsche, U., Iriarte, L., 2014. Sustainability of solid wood bioenergy feedstock supply chains: operational and international policy perspectives: IEA Bioenergy Task 40. Available from: http://www.bioenergytrade.org/downloads/t40-sustainable-wood-energy-2014.pdf

Thiffault, E., Endres, J., McCubbins, J.S., Junginger, M., Lorente, M., Fritsche, U., Iriarte, L., 2015. Sustainability of forest bioenergy feedstock supply chains: local, national and international policy perspectives. Biofuel. Bioprod. Bioref. 9 (3), 283–292.

USFS, 2011. National Report on Sustainable Forests—2010. United States Forest Service, Washington, DC, Available from: http://www.fs.fed.us/research/sustain/docs/national-reports/2010/2010-sustainability-report.pdf.

Zanchi, G., Pena, N., Bird, N., 2010. The Upfront Carbon Debt of Bioenergy. Joanneum, Graz, Austria, Available from: http://www.transportenvironment.org/publications/upfront-carbon-debt-bioenergy.

Chapter 9

Constraints and Success Factors for Woody Biomass Energy Systems in Two Countries with Minimal Bioenergy Sectors

David C. Coote*,†, Evelyne Thiffault, Mark Brown†**

*School of Ecosystem and Forest Sciences, University of Melbourne, Richmond, VIC, Australia;
**Department of Wood and Forest Sciences and Research Centre on Renewable Materials, Laval University, Quebec City, Canada; †Australian Forest Operations Research Alliance, University of the Sunshine Coast, Sippy Downs, QLD, Australia

Highlights

- Australia and Canada have extensive fossil fuel deposits, with no significant concerns expressed in national government level policy regarding energy security risks and, as of 2015, relatively slow adoption of material climate change mitigation policies.
- Fossil fuels comprise a large percentage of Australian and Canadian exports and both economies are more energy-dependent than any of the European nations compared in this study.
- Canadian wood pellet exports aside, there have been some success stories on the Australian and Canadian woody biomass energy front, despite the lack of strong economic, policy and energy security drivers to encourage the development of a renewable energy sector.
- Niche applications of forest bioenergy, through initial deployment of small-scale solutions, provide a platform for supply chain development. When better integration and equipment is introduced and local stakeholders and communities increase their understanding of bioenergy, significant efficiency gains are revealed that facilitate the establishment and expansion of bioenergy opportunities.
- Technology learning-through-doing in niche applications, leveraged with knowledge from the best practices and experience of well-established bioenergy supply chains from other countries, represent an opportunity to set strong foundations for bioenergy in countries with immature bioenergy sectors.

165

INTRODUCTION

Processing woody biomass into energy, using modern systems with feedstock from well-developed biomass supply chains, is common in a number of countries from boreal and temperate biomes (IEA, 2014a) (for examples of supply chains, see chapter: Comparison of Forest Biomass Supply Chains from the Boreal and Temperate Biomes). Energy from woody biomass is an important component of the primary energy supply in many countries (IEA, 2014a) and, in some circumstances, can be competitive with energy from fossil fuels (ECOFYS, 2014). While some of these countries may further develop their domestic woody biomass energy sector to contribute to proposed global bioenergy targets, achieving these targets would be assisted by developing new supply chains for energy biomass in other countries. Development of a nascent biomass energy sector will inevitably start on a small scale, with only nominal contributions to global or national targets; even countries with large biomass sectors had only niche deployments of modern woody biomass energy systems, at some point. Biomass supply chains can be used to supply domestic energy systems, or can be exported to countries with requirements for biomass that can't be satisfied by domestic sources. Where biomass exports are possible under a particular set of economic conditions, as discussed in chapter: Challenges and Opportunities for International Trade in Forest Biomass, the export market may depend on subsidies and policies in importing nations that could change in the future. Hence, a country such as Canada, exporting around 1.3–1.6 million tonnes of wood pellets per year to Europe (see chapter: Challenges and Opportunities for International Trade in Forest Biomass) but with comparatively few modern bioenergy plants for local use, may find it advantageous to develop internal energy biomass markets to complement European export markets. This would contribute to the development of local bioenergy expertise and stabilise biomass supply chains against external policy changes, thereby improving the stability and profitability of the biomass sector and, ultimately, benefiting international trade (Thrän et al., 2014).

A supply chain for the woody biomass used in energy systems comprises several stages, including collection and transportation to an energy conversion plant where heat and/or electricity will be produced (see chapter: Challenges and Opportunities of Logistics and Economics of Forest Biomass). When processes such as fast pyrolysis reach commercial status, liquid fuels will represent another potential output (see chapter: Challenges and Opportunities for the Conversion Technologies Used to Make Forest Bioenergy). Biomass supply chains may include intermediate steps, such as chipping, pelletisation and torrefaction. In some cases (such as sawmills and pulpmills), residual woody materials could be processed in situ using co-located energy conversion plants that would collapse the supply chain by simply moving biomass around the plant site. Countries such as Austria, Germany, Finland and Sweden have developed sophisticated supply chains based on collecting, for example, forest harvest residues and transporting these materials to energy plants (for supply chain examples from several countries, see chapter: Comparison of Forest Biomass Supply

Chains From the Boreal and Temperate Biomes). These supply chains may use multiple transportation modes and utilise intermediate storage sites, thus shifting the availability of biomass both temporally and spatially.

A crucial point in any energy-related analysis is that society, as a whole, is largely uninterested in energy per se, but does demand access to services such as space heating, cooling and artificial lighting; electricity to power equipment in residential, commercial, industrial and government sites; thermal energy for industrial and agricultural activities; and liquid fuels for transportation. Within different temporal and spatial contexts, the form and scale of the energy available from a particular technology will differ in relevance and utility. When considering an increase in the fraction of renewables in the energy generation mix, a number of factors must be considered. These include the levellised cost of energy from new plants compared to existing and new fossil fuel plants, the cost of developing transmission/distribution infrastructure (ECOFYS, 2014) and how costs and benefits will be shared among players within the energy sector, for example through a mix of subsidies (Badcock and Lenzen, 2010), feed-in tariffs, generation quotas and other mechanisms (Haas et al., 2011). Of increasing concern is anthropogenic climate change, caused by elevated levels of atmospheric greenhouse gases (GHG) and risks to energy security, caused by depletion of local fossil fuel resources and/or geopolitical tensions. Using renewable energy such as biomass can reduce net GHG emissions and improve local energy resilience.

Societies will differ on how they value the range of factors influencing the energy sector and, in particular, energy from woody biomass (see the discussion on social aspects of bioenergy in chapter: Economic and Social Barriers Affecting Forest Bioenergy Mobilisation: A Review of the Literature). A full analysis would require considerable cross-disciplinary resources in a number of diverse fields and exceed the scope of this chapter. However, it is illustrative to identify the characteristics of individual countries that have fostered the development of a woody biomass industry (either for local use or for export) and compare these with countries that lack a significant biomass energy sector. In this chapter, we discuss relevant characteristics of several countries where energy biomass supply chains have been successfully developed and contrast these with countries where energy biomass systems are less mature. We also identify some examples of woody biomass energy systems that have been successful in countries that have neither a broad biomass energy sector nor strong policies to support the bioenergy sector. Finally, we examine the potential for further expansion of these systems.

ANALYSIS

Comparing Austria, Finland, Germany and Sweden with Australia and Canada

We highlight illustrative differences between Austria, Finland, Germany and Sweden—countries with well-researched, successful woody biomass energy sectors—and Australia and Canada. Björheden (2006) proposed a list of possible drivers for the uptake of woody biomass for energy generation in modern combustion plants. As well as techno-economic capabilities, such as suitable wood harvesting machinery, Björheden (2006) also included such factors as energy security, GHG mitigation, concerns about nuclear power and legislative measures.

National Forest Sectors

At any scale, be it local woodlots that produce firewood or large energy plants that consume forest residues, the forest sector is the start of the woody biomass supply chain. All six of the countries included in this analysis have active forest sectors (FAO, 2010). Forest management practices in Austria, Finland, Germany and Sweden support a high rate of industrial roundwood production relative to their forested areas. By comparison, Canada has the largest forested landbase among the six countries, but a high proportion of this landbase is considered non-commercial, for economical or ecological reasons. Harvesting timber from the native forest estate in Australia is contentious and the subject of frequent community debate. Moreover, relative to the size of their forested landbases Canada and Australia remove only small quantities of woody biomass (see chapter: Quantifying Forest Biomass Mobilisation Potential in the Boreal and Temperate Biome). Nevertheless, forest products represent a large share of the total exports from Finland, Sweden and Canada.

Energy Sectors

Other than coal (mostly lignite) deposits in Germany, Austria, Finland, Germany and Sweden have limited domestic fossil fuel resources (Table 9.1). By contrast,

TABLE 9.1 Domestic Coal, Oil, Natural Gas, Hydro-Energy and Uranium Resources in 2012 (BP, 2013; IEA, 2014b; OECD, 2010)

Country	Austria	Finland	Germany	Sweden	Australia	Canada
Coal (10^9 tonnes)	0	0	41	0	76	7
Oil (thousand million barrels)	0	0	0	0	4	174
Natural gas (trillion cubic metres)	0	0	0	0	4	2
Hydro (GW installed)	14	3	11	16	9	76
Uranium (tonnes U, recoverable as at 1/1/2009, <US$80 per kg)	0	0	0	0	1,612,000	447,400

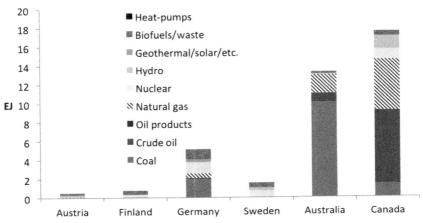

FIGURE 9.1   Production of primary energy from domestic sources in 2012 (IEA, 2014c).

Australia and Canada have large deposits of coal and other fossil fuels, as well as massive uranium deposits. In addition, all six nations, particularly Canada, have developed hydro generation capacity.

According to IEA methodology, total primary energy production is the sum of domestic energy production and the net of imports, exports, aviation and marine bunker volumes and stock changes (IEA, 2014c). Australia and Canada have much larger domestic primary energy production than the European countries (Fig. 9.1). Fossil fuels dominate domestic primary energy production in Australia and Canada, with Canada also producing 5.9% of its primary energy from nuclear power and 7.8% from hydro (Fig. 9.2). The European nations tend to produce much lower quantities of domestic primary energy (Fig. 9.1), with only small contributions from fossil fuels (Fig. 9.2). One exception is Germany,

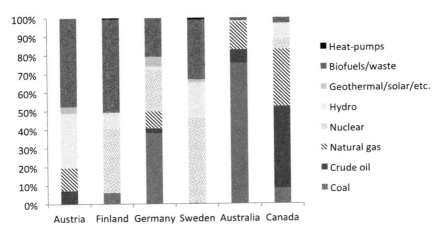

FIGURE 9.2   Composition of primary energy from domestic sources in 2012 (IEA, 2014c).

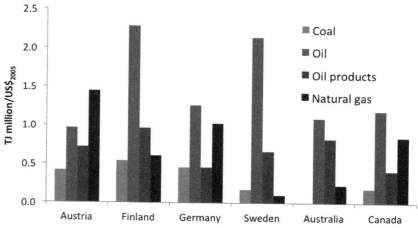

FIGURE 9.3 Energy imports, normalised for GDP, in 2012 (IEA, 2014b).

which relies on coal for nearly 40% of its domestic energy production. Biomass is a large proportion of domestic primary energy production in the European nations, as is nuclear (except in Austria) and hydro.

Energy security became a major issue in Europe during the first oil shock in the early 1970s. The shortage of fuel for transportation and oil for space heating had widespread direct and indirect impacts on commercial, industrial, government and domestic customers. More recently, Russia has periodically intervened in the natural gas supply to Europe, causing further energy security concerns (EC, 2014; Le Coq and Paltseva, 2009). Other instances of social unrest, revolution and war could also disrupt oil supply. All four European nations import a substantial amount of the energy they consume (Fig. 9.3) and have a limited ability to produce primary energy from domestic sources (Fig. 9.4). Because of a decline in the productivity of Australia's oil fields and continuing closure of local refining capacity, Australia imports oil and oil products at higher quantities per normalised GDP unit than some of the European nations. Other than this dependency, however, Australia and Canada import little energy and have high energy security for the other fuels presented.

The large domestic fossil fuel resources in Canada and Australia support strong fossil fuel sectors, with active development of new projects and large exports (Fig. 9.5). There are no significant concerns expressed by either national government about sovereign energy security, although there are large regional differences within the two countries. Oil reserves and domestic liquid fuel refining capacity have both declined considerably in Australia since their peak several decades ago. In Canada, energy has long played a central role in the growth of the economy. In recent years, Canada has become increasingly reliant on direct investment for the development of domestic oil and gas reserves (Krupa, 2012).

In general, household and industry prices for natural gas and electricity are higher in European countries than in Canada (Fig. 9.6). One exception is the

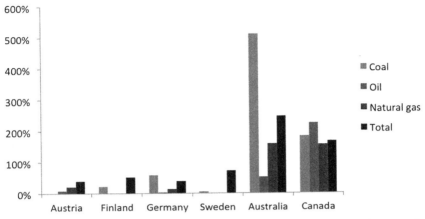

FIGURE 9.4   Proportion of national primary energy production met by domestic sources in 2012 (IEA, 2014b) .

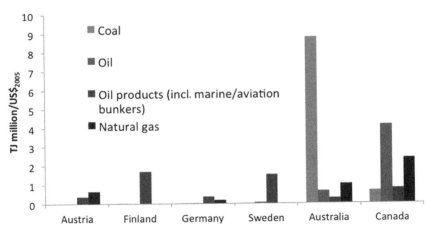

FIGURE 9.5   Exports of energy, normalised for GDP, in 2012 (IEA, 2014b).

price of electricity for industrial use, which is comparable in Canada, Finland and Sweden [Australian industry and household energy prices are not presented in recent analyses (IEA, 2014d)].

   In Sweden, the mix of primary energy sources used to generate electricity on its own [ie without combined heat and power (CHP) generation] is based almost entirely on non-fossil fuel sources (Fig. 9.7). This contrasts sharply with Australia, where electricity is generated primarily from fossil fuels, with a small contribution from hydro and non-hydro renewables (wind, solar and biomass). Germany produces a large fraction of its electricity from non-hydro renewables, but also remains dependent on coal, while electricity production in Finland and

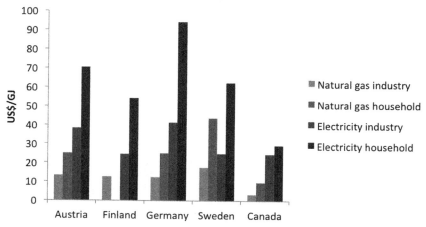

FIGURE 9.6 Gas and electricity prices in 2012 (IEA, 2014d).

Austria is less dependent on fossil fuels. Fig. 9.7 also shows that the Canadian and Australian economies, with their high ratio of electricity generation per unit of GDP, are much more dependent on electricity than any of the European nations presented.

Of particular relevance to the woody biomass energy sector in Austria, Finland, Germany and Sweden are government policies promoting CHP systems and other commercial heat applications supplying, for example, district heating systems (CEER, 2015). Biomass thermal and co-generation systems can supply thermal energy to a district heating system (DHS) or co-located thermal demand system, improving the economics of a biomass energy conversion system (ECOFYS, 2014). Obernberger and Thek (2008) suggest that decentralised biomass CHP systems should be based around thermal demand and operate in

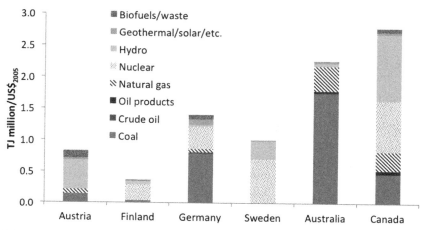

FIGURE 9.7 Electricity generation mix in 2012, normalised for GDP (IEA, 2014c) .

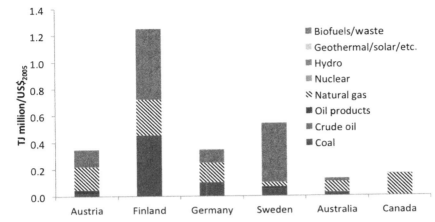

FIGURE 9.8   Fuel used for commercial CHP in 2012, normalised for GDP (IEA, 2014c).

heat-controlled mode. Commercial CHP and heating systems are much larger per normalised GDP unit in the European countries than in Canada and Australia (Figs 9.8 and 9.9). Biofuels/waste are the most significant fuel for CHP systems in Finland and Sweden and also dominate commercial heat plant generation in Austria and Sweden. As an interesting indicator of potential future energy paths, Austria also produces some commercial heat from other low-carbon sources.

By contrast, there is very little use of biofuels/waste for commercial heating systems in Canada and Australia. Indeed, Australia and Canada generate insignificant quantities of what the IEA classifies as commercial heat (Fig. 9.9). DHS are almost non-existent in Australia beyond campus-scale systems. There are

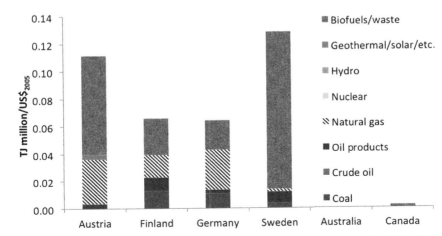

FIGURE 9.9   Fuel used for commercial heat production in 2012, normalised for GDP (IEA, 2014c).

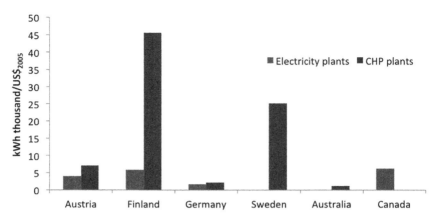

FIGURE 9.10   Use of commercial woody biomass energy for electricity generation in 2012 (IEA, 2014a).

also relatively few DHS systems in Canada (CIEEDAC, 2014), indicating lower uptake of DHS in Canada than in Europe. The European nations also use far more wood in CHP and heating systems than Canada and Australia (Figs 9.10 and 9.11).

While renewables such as wind and solar generate energy intermittently, woody biomass systems are dispatchable (ie energy output is available when required). Woody biomass energy systems produce significantly lower GHG external costs than fossil fuels. As well as the benefits of dispatchability, the levellised cost of energy (LCOE) production from woody biomass systems can be competitive with other renewables depending on a number of factors. For example, all four of the European countries considered in this assessment

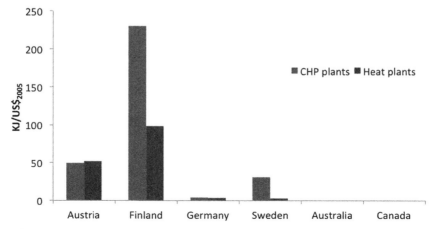

FIGURE 9.11   Use of commercial woody biomass energy for heat generation in 2012 (IEA, 2014d).

experience lower levels of insolation (ie the amount of solar radiation energy received) than the EU-28 average; low levels of insolation increases the LCOE from photovoltaics (PV) (ECOFYS, 2014). On the other hand, much of Australia receives very high levels of insolation (IRENA, 2016), making solar power generation considerably more competitive. Extensive capacity for generation of hydroelectricity has also been installed in several Canadian provinces and commercial wind power generation exists in some parts of Canada and Australia (IRENA, 2016).

Government Support

Austria, Finland, Germany and Sweden all have significant renewable energy programs that include woody biomass. For example, the German Government's Energy Policy Road Map (BMU, 2011) lists a number of steps for increasing the production of renewable energy by 2020. These include doubling the contribution of renewables (including biomass) to electric and thermal energy production, increasing the use of biofuels and doubling the use of CHP generation. To this end, the Renewable Energy Sources Act (EEG) and Electricity Feed Act are credited with facilitating the substantial growth in renewable electricity observed to date. The EEG was amended in 2009 to provide selective investment incentives for bioenergy. The Road Map further mentions (BMU, 2011, p 22) that while 'there is huge potential [...] for [...] using more renewable sources, not enough attention has hitherto been paid to the heat market' and lists a number of existing legislative measures to reduce GHG emissions from thermal energy, including the CHP Act and the Renewable Energies Heat Act. According to the Road Map, the German government aims to expand District Heating Networks, decentralise energy supply from smaller scale CHP systems and industrial cogeneration and expand the number of biomass CHP plants (Thrän et al., 2012). Under this broad program of renewable energy development (often referred to in Germany as the *Energiewende*) Germany, a major industrial power, managed to reduce national GHG emissions (excluding land-use, land-use change and forestry, ie LULUCF) by 25% between 1990 and 2012 (Fig. 9.12).

Stakeholders generally state that the development of a biomass sector that uses modern combustion technology requires some degree of government support. This support can take various forms, including direct subsidies, technical and financial assistance to establish energy plants and operational subsidies to maintain them (ECOFYS, 2014; EEA, 2014). Government funding for university and industrial research and development can assist in the development of new technologies, help improve the efficiency of existing technologies, measure social and community costs and benefits, identify where market failures exist and determine how such failures can be addressed. With the help of such government support, Austria's total emissions (excluding LULUCF) have stayed flat since 1990, while other European nations have reported substantial reductions in total GHG emissions (Fig. 9.12). By contrast, Australia and Canada

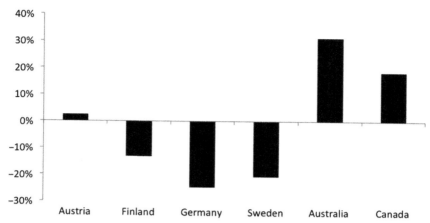

FIGURE 9.12 Change in national GHG emissions (%), 1990–2012 (excluding LULUCF) (UNFCCC, 2014).

have increased total emissions (excluding LULUCF) over this period. GDP-normalised emissions have dropped substantially between 1990 and 2012 in the European nations (Fig. 9.13), with smaller declines in Australia and Canada. Emissions per primary energy unit also fell substantially between 1990 and 2012 (Fig. 9.13) in the European nations, while the GHG emissions intensity of primary energy production in Australia and Canada showed little change.

European Union nations offer a range of subsidies and feed-in tariffs for heat and electricity production from biomass (CEER, 2015). Indeed, all four of the EU countries considered in this analysis have had subsidy programs for renewables, including biomass, for decades (Haas et al., 2011). The Upper Austrian Energy Agency refers to a co-ordinated campaign of 'carrots, sticks, and tambourines' using incentives, regulatory measures and educational outreach (Egger et al., 2010). Bioenergy thermal plants supplying heat for DHS at the community scale are common. Austrian communities considering the installation of a biomass energy system can also access extensive government assistance with planning, installation and operation.

As of 2015, there is no co-ordinated strategy to significantly expand the use of woody biomass for stationary energy generation in Australia. Some initiatives have targeted bioliquids, but without the benefit of any demonstration plants or commercial deployments. There is essentially no government support aimed specifically at the development of woody biomass-based thermal systems in Australia. Rather than encouraging this sector, subsidies are available in some areas to replace wood heaters with natural gas heaters. There is no Australian equivalent of the UK Renewable Heating Incentive or similar schemes offered in Europe (CEER, 2015). Electricity generated from woody biomass from the native forest estate was included in the primary Australian renewables support program, the Renewable Energy (Electricity) Act 2000 (Cth) (Australia, 2000)

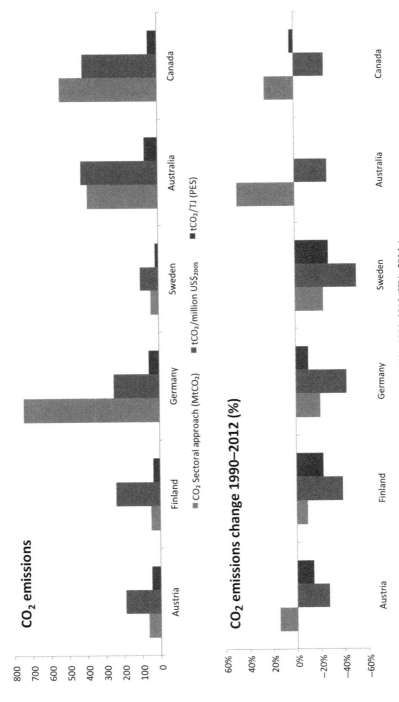

FIGURE 9.13 GHG emissions in 2012 and change in GHG emissions (%), 1990–2012 (IEA, 2014e).

(with very tight regulatory control of eligibility under the Renewable Energy (Electricity) Regulations 2001 (Cth) (Australia, 2001). However, all native forest woody biomass, including harvest and processing residues, was specifically excluded in 2011 (Australia, 2011). Eligibility for some native forest biomass was restored in 2015, with the Renewable Energy (Electricity) Amendment Bill 2015 (Cth) (Australia, 2015). Eligibility of wood from plantations is also under tight regulatory control. Other than an exception for some timber processing residues, burning wood from State-owned native forest estate for electricity generation in systems larger than 200 kW is illegal in the state of New South Wales (NSW) (NSWGOV, 2013). With some restrictions, however, residues from NSW private native forests may be burned in energy conversion systems. In general, there are no subsidies in Australia for renewable heat, other than for solar hot water. Interestingly, heat pumps can receive subsidy support originally provided for renewables, presumably on the basis that they use GHG-intensive energy from the Australian grid more efficiently than resistive heating methods. The state of Victoria also allows feed-in tariffs—originally aimed to encourage renewables—for 'low-emission' technologies (VICGOV, 2013).

There is also no nationally coordinated bioenergy strategy in Canada, nor are there specific national targets for biomass development. In 2013, the Canadian National Energy Board released forecasts of energy sector demand and composition to 2035 (NEB, 2013). The proportion of stationary energy derived from biomass is not forecast to change significantly during this period. Relative to the initiatives encouraging fossil fuel development, comparatively minor incentives have been offered at the federal level for renewable energy deployment (Krupa, 2012) and favourable treatment of fossil fuels has been identified as an important barrier to market penetration of renewable energy (Painuly, 2001). Nevertheless, initiatives at the provincial level have facilitated the deployment of several bioenergy projects, including the conversion of coal power plants to biomass in the province of Ontario and the installation of community DHS, using biomass to replace heavy oil, in the provinces of Quebec and British Columbia (BC). Policy support for forest bioenergy development is often embedded in larger government efforts to revitalise the forest sector, which has been struggling with the fall in demand for paper and the economic downturn in the United States. Another policy driver is the need to assist specific regions dealing with catastrophic natural disturbances, such as the mountain pine beetle epidemic in British Columbia's forests. One outcome of these efforts has been an increase in wood pellet production capacity, which is almost entirely directed towards international export because of a lack of internal demand and low expectations for the growth of the domestic market.

Socio-Political Factors

Austria, Finland, Germany and Sweden are all signatories under EU climate change mitigation initiatives. All four countries have active renewable energy programs. Of the large industrial economies, Germany has arguably the most aggressive renewable energy program under the range of initiatives known as the *Energiewende*. The state of Upper Austria has targeted 100% renewable energy for space heating by 2030 (Egger and Öhlinger, 2009). Although Finland, Germany and Sweden have nuclear power generators, public opposition to nuclear power is widespread in Austria and Germany and is growing in Nordic countries. Following the release of radioactive material from the Fukushima nuclear plant as the result of an earthquake in 2010, the German government announced in 2011 that existing nuclear plants will be closed and no new plants will be built. Subsidy programs and government support for bioenergy project developments assist regional and community acceptance of woody biomass systems. With centralised energy systems, energy is imported into a region or community and money is exported. Cost-effective, decentralised renewable energy systems can ensure that money spent on energy is retained within the community. Where surplus generation exists, energy can be exported for financial return. Studies such as those by Hirschl et al. (2010) illustrate the value chain and local economic benefits from woody biomass systems and other renewables. Improved local energy resilience is another benefit of these systems.

Pollutant emissions from biomass combustion systems—notably particulate matter—are recognised as a less appealing characteristic of this technology and can cause community concern. EU directives and, in some countries, national legislation regulate the levels of pollutants, including particulate matter, generated from biomass combustion systems. Recent policy activity in this area has seen moves toward stricter EU-wide regulation of woody biomass energy systems at a range of scales. A number of European initiatives are also working on reducing pollutant emissions from new biomass systems to levels a small fraction of existing commercial systems.

In Australia and Canada, however, the need for climate change mitigation remains a controversial issue. As of mid-2015, there is relatively little political support (especially at the federal level) for near-term, large-scale reductions in anthropogenic GHG emissions. By contrast, Sweden has had a long-standing GHG tax. Australia introduced a carbon tax in 2012 and removed it in 2014, following a change in government. Current Australian policy is a 5% reduction in GHG emissions by 2020, based on 2000 as the baseline year. Nuclear fission is considered a zero GHG emissions technology in Australian Government publications (BREE, 2012) and receives attention as a possible component of a more aggressive policy response to GHG reductions. If policy is changed to allow electricity generation from nuclear fission, Australia has a large fraction of the world's currently identified commercially extractable uranium ore (Table 9.1). Indeed, several uranium mines already operate in Australia. Under the Copenhagen Accord, Canada has committed to reducing its GHG emissions by 17% from 2005 levels, by 2020. The government has contributed to projected emission reductions by regulating GHG emissions from the transportation and electricity sectors, but, as of 2015, is still discussing ways to reduce emissions from other sectors, including oil and gas. Canada has a low GHG emissions grid, due to extensive nuclear and hydro generation capacity. Nevertheless, substantial fractions of GDP in both Canada and Australia stem directly from energy extraction industries and the competitive advantage, incomes and livelihoods of large regions depend on them. Consequently, the incentives for replacing the current energy mix with renewables are not as strong as those in European countries.

There is also substantial opposition to native forest harvesting and the use of forest residues for energy production in Australia. Environmental groups on both sides of the Atlantic have raised concerns about the harvesting of Canadian forests for bioenergy production (especially for wood pellet export), because of potential threats to boreal forest ecosystems. Whether or not these criticisms are valid, they represent a constant source of opposition to the development of forest biomass energy supply chains.

Local Expertise

Woody biomass systems in Austria, Finland, Germany and Sweden represent the outcome of a long-term, coordinated strategy between industry, government and academia. As a consequence, these countries have well-established expertise in the development of woody biomass energy system supply chains, finances, project management, system configuration, installation and operation. In Australia and Canada, there is only limited expertise in the commercial deployment of modern woody biomass energy systems. Research and other initiatives have tended to focus on technologies, such as second-generation liquid biofuels from pyrolysis and gasification of woody biomass that have yet to reach large-scale commercial deployment, creating market uncertainties. There is limited local expertise with direct experience in these systems, which hinders informed assessment of potential biomass energy projects. And as companies involved in installations are unfamiliar with these systems they may charge margins containing higher risk premiums to cover unexpected costs. There is also limited experience with woody biomass energy systems at the community scale in Canada and Australia. Although communities contemplating the installation of biomass energy systems in many parts of Europe can easily visit an operational plant, demonstration sites in Australia and Canada are limited.

Successful Niche Applications of Bioenergy Systems in Canada and Australia

Even without significant drivers for the installation of woody biomass energy systems in Canada and Australia, however, there are a number of recent success stories. Most woody biomass systems in these two countries are either part of forest sector processing sites, such as sawmills and pulp plants, or are located close to these sources of material. There has been no recent inclusive research covering economic and other drivers of these plants, but industry sources suggest that these systems are typically established to provide less expensive electricity and thermal energy for onsite use and to reduce the costs incurred in processing waste residues. Upstream diversion of clean wood to energy plants, including co-location of biomass energy systems within plants that produce woody residues, provides an easily accessible local supply of biomass. However, it is still legal in Canada and Australia to landfill biogenic material (although it can be forbidden at the provincial or local level). Although several Australian states have increased levies to encourage landfill diversion, there are no incinerators or 'waste-to-energy' plants in Australia and no organised programmes to divert clean wood for energy production.

Almost all of the forest bioenergy used in the Canadian industrial sector comes from residues generated by the forest industry, especially pulp and paper. In 2009, forest biomass (mainly from processing residues, such as sawdust and black liquor) accounted for 58% of the total energy used by the forest industry as a whole and 62% of the total energy used by the pulp and paper sector (NRCAN, 2011) (see also chapter: Challenges and Opportunities for the Conversion Technologies Used to Make Forest Bioenergy). The forest biorefinery concept has also received substantial interest from the pulp and paper sector. The Canadian forest sector has been in a precarious economic situation for some time. A decline in traditional manufacturing has driven interest in using renewable resources such as forest biomass in alternative ways to develop sustainable and profitable businesses (Moshkelani et al., 2013). A large number of forest biorefinery approaches and strategies have been proposed across Canada and a screening process has already begun to identify and implement the most cost-effective of these options. Examples include dissolving pulp, extracting hemicellulose from wood chips before pulping, recovering lignin from black liquor, recovering methanol from evaporator or digester condensates and producing biogas from sulphite mill condensates or biological sludge (Paleologou et al., 2011).

Modern woody biomass energy systems in Australia include kilns at sawmills and biomass thermal systems at other Australian businesses such as abattoirs, several swimming pools and a hospital with thermal energy demand and access to waste woody biomass from sawmills. There are several CHP plants at pulpmills, two small Organic Rankine Cycle (ORC) electricity generating systems at sawmills and some bagasse CHP plants can burn woody biomass.

One example is the Reid Brothers sawmill in Yarra Junction, east of Melbourne, which installed a modern 1 MW_{th} boiler to provide heat for the hardwood kilns. Previously, the sawmill used liquefied petroleum gas (LPG) to generate thermal energy for the kilns and disposed of the waste generated from milling the timber. The LPG cost was increasing as was the cost of removing the sawmill waste. Burning the waste solved both of these problems. The sawmill operators estimate a simple payback period of several years. More recently, a small ORC system developed and operated by an Australian company has been installed at the sawmill. The ORC system can use steam generation capacity in excess to what is required by the wood drying process. There is little marginal cost in producing steam used by the ORC, as the wood waste generated onsite is free, with few other profitable markets. The sawmill and ORC operator have an off-take arrangement where the sawmill is charged for electricity generated by the ORC and consumed onsite. This arrangement reduces the operational risk for the sawmill operator. The $1MW_{th}$ system has operated successfully for over 10 years. This system had a well-defined favourable economic basis, little risk to feedstock supply, as long as the sawmill is in operation and used a modern, highly automated combustion system. There is also a favourable GHG emissions reduction outcome by substituting wood for the LPG previously used.

In Canada, access to reliable energy supply can be constrained in remote, off-grid communities (there were 292 off-grid communities across Canada in 2011) (AANDC-NRCAN, 2011). Most of these communities use diesel-powered generators to produce electricity, but fossil fuels are particularly expensive and can result in spills. Little has been done to integrate local resources into the energy mix in these remote communities (AANDC-NRCAN, 2011). More than 175 First Nations communities across the country meet some of their electricity needs using diesel generators and 140 rely solely on diesel (Arriaga et al., 2013). However, there is a growing interest in capitalising on local sources of woody biomass to generate electricity.

The Cree community of Oujé-Bougoumou, in northern Quebec, has had a long history of resettlement caused by resource exploration in its territories. In 1992, the 600-member community reached an agreement with the province of Quebec and the federal government to ensure financial support for the construction of a new permanent village in Nord-du-Québec, a region with a thriving forest sector. Although this region was connected to the provincial power grid, the community decided to install a forest biomass-based DHS because of its potential for controlling energy costs and for contributing to community development. Since installation, feedstock has been provided primarily by residues from a sawmill located 30 km away. However, biomass sourcing directly from the forest is also being considered. Despite several operational challenges (eg a lack of local expertise in equipment maintenance), the district heating system in Oujé-Bougoumou is recognised as one of the few successful community bioenergy projects in Canada. Several factors have distinguished this case from similar projects; these include strong leadership, widespread community support and reliable workers. The system is also strongly linked to the First Nations community's objective of self-sufficiency. Furthermore, the community is highly motivated to provide modern, efficient and affordable housing and the economic benefits of the DHS are re-invested in the housing program. Therefore, access to a reliable supply of heat and electricity, such as that provided by forest biomass-based systems, can be of significant benefit to remote communities, particularly First Nations ones. According to Krupa (2012), further deployment of First Nations-based renewable energy projects could be encouraged through the use of price incentives for electricity generation and transmission, additional funding and financing bodies dedicated to First Nations projects and continuing education programs that incentivise on-reserve First Nations to build project development capacity. The high cost of electricity in remote communities (both First Nations and non-First Nations) is a significant deterrent to economic development opportunities and adds to the cost of living for populations who often live at a subsistence level. Meanwhile, many remote Canadian communities have access to adequate renewable energy resources to meet both power generation and space heating applications. Appropriate use of these resources could contribute to sustainable development in these communities. The economic benefits stemming from the use of local sustainable resources in these communities could include job creation, local skill development and increased community self-reliance (AANDC-NRCAN, 2011).

Another niche application of forest biomass systems in Canada is that of the mining industry. Although limited in scope and time, occasional fossil fuel shortages can disrupt mining activities, which have very tight profit margins. Moreover, there is a growing interest among mining industry representatives in improving sustainability, encouraging community involvement and integrating with other regional stakeholders. Mining and forestry often occur in the same regions, providing the potential for collaboration between the two sectors. In Abitibi-Témiscamingue, in north-western Quebec, for example, fossil fuel shortages have encouraged a gold mine company to look for more stable alternative (or back-up) sources of energy and has therefore installed a biomass-based heating system for its underground mining installations (Boileau, 2015).

Metallurgical industries (eg steel and metallic silica plants), which require large amounts of heat and carbon as a reducing agent, could also develop niche applications for forest biomass. The use of wood charcoal (or other wood-based products) in these industries is very rare in Canada, although it is more common in developing countries. However, the abundance of forest residues and the presence of well-organised networks of forest-based communities could encourage the use of woody biomass in Canadian metallurgical plants (Dessureault, 2015). In order to develop woody biomass energy plants in collaboration with the mining or metallurgical sectors, partnerships with local forest companies or cooperatives must be fostered. As well, development of local expertise and knowledge on biomass supply chains will also be required to ensure that much larger volumes of biomass can be mobilised within the region than are required for small DHS. Similar interfaces between the industrial and forest sectors have been pursued in other countries within the boreal and temperate biome that have large manufacturing sectors, for example Finland. In Australia, small quantities of locally produced carbon from woody biomass are used in smelting silicon and activated carbon is imported for use in purifying gold.

Challenges and Opportunities of Niche Applications in Countries Without Broadly Successful Woody Biomass Energy Sectors

Biomass energy systems that are not situated at a timber processing facility require a supply chain. In immature biomass energy markets, such as those in

Canada and Australia, the supply chains tend to be ad hoc in nature. Without an existing industry, biomass supply chains are often established and trialled as new and independent operations using available forestry and/or agricultural equipment. Inefficiencies resulting from the use of equipment not ideally designed or sized to the task, compounded with a lack of integration with existing supply chains for other products means that these ad hoc trials can easily exceed cost expectations. As a result, trial projects are often limited in scope and scale, in order to minimise the length of supply chains and, thus, their cost. Such small scale trials also prevent stakeholders from developing the specific expertise and technological learning required to achieve agility and cost efficiencies along the supply chain. Moreover, social acceptability of projects is more difficult to acquire: bioenergy projects can be perceived with suspicion by the public, who may have concerns for the environment, for human health and for local autonomy (Eaton et al., 2014). Nevertheless, these early, small-scale trials present the platform for supply chain development. When better integration and equipment is introduced and local stakeholder and community buy-in is achieved, significant efficiency gains can facilitate the expansion of existing bioenergy systems and the establishment of new opportunities.

Short of major government initiatives, a number of factors may encourage the development of the biomass energy sector in Australia and Canada. In Australia, biomass thermal systems with sufficient operational hours and low-priced feedstocks are already competitive with thermal energy from LPG and reticulated natural gas. The cost of gas in Australia may increase to world parity, after large export Liquefied Natural Gas (LNG) plants on the East Coast begin operation. If gas prices do rise, biomass thermal energy will become cost-effective on a wider range of sites. Electricity prices have also increased considerably in Australia; the electricity index used by the Australian Bureau of Statistics to calculate the Consumer Price Index doubled between 2006 and 2014 (ABS, 2014). In recent years, the establishment of simplified connections to the grid, as well as feed-in tariffs, have resulted in the installation of nearly 4 GW of rooftop PV, spread over 1.4 million sites, by the end of 2014 (CER, 2015). Although feed-in tariffs have declined, biomass electricity and CHP plants could also sell electricity surplus to the grid if regulatory policy continues to allow embedded and decentralised renewable energy generators to connect to the grid at a tolerable cost.

Accurately forecasting demand for stationary energy over a longer timeframe is difficult. In Canada, aging electricity generation capacity could be refurbished, replaced by new facilities using the same fuel mix, replaced by new facilities using nuclear, fossil fuel, hydro or renewables, or made unnecessary by increased emphasis on energy conservation. The addition of new generation capacity represents an opportunity to transition to renewable energy sources. Kedron (2015) argues that both geography, as well as regional, national and international policies contribute to the creation of market niches for biofuels. New technologies can be trialled in the regions to which they are best suited.

New renewable generators that use woody biomass, small-scale solar or hydro can be embedded closer to energy demand. Woody biomass CHP and thermal energy systems can be used to meet industrial, commercial, government, community and residential energy requirements. The 2013 District Energy Inventory for Canada identified 116 operating district energy systems in Canada (CIEED-AC, 2014), with the majority in Ontario and in First Nations communities. After natural gas, biomass was the second most popular base load fuel reported for Canadian district energy systems. BC and Ontario are both implementing feed-in tariffs for renewable energy. Feed-in tariffs are perceived by investors in both North America and Europe as the most effective renewable energy policy (Bürer and Wüstenhagen, 2009). The province of Ontario, with a population of 13 million, has recently announced a number of initiatives for increasing non-hydro renewables and reducing coal-powered energy generation. Furthermore, the Green Energy and Green Economy Act of 2009 has imposed feed-in tariffs for wind, solar PV, bioenergy and hydropower (Stokes, 2013). Over the past decade, non-hydro renewables have grown to approximately 2500 MW. In 2010, the Ontario government proposed a target of 13% non-hydro renewable energy capacity by 2030, with generation capacity of 10,700 MW from wind, solar and bioenergy by 2018 (Stokes, 2013). The province of British Columbia's Energy Plan has allowed the production of more than 1600 GWh per year of new bioenergy power, facilitated access to wood fibre for energy use via changes to the BC Forest Act and has started to bring bioenergy capacity to remote and First Nations communities (BCBE, 2011).

Studies such as the Australian Energy Market Operator 100% Renewable Australia (AEMO, 2013) have modelled contributions from biomass to total electricity generation of up to 18% from distributed Steam Rankine Cycle generators. Distributed biomass systems increase local energy resilience and benefit local economies. By generating energy locally, communities can retain money that would otherwise be exported (Hillring, 2002; Hoffmann, 2009) and, potentially, make money from energy exports. Bioenergy villages in Germany have illustrated this potential (BEL, 2016). Forest bioenergy production from supply chains that are located entirely within the community or region containing the bioenergy plant can reduce exposure to interruptions of energy supply from centralised systems. In some remote areas of Quebec, supplies of propane or heavy oil can be interrupted for weeks, leading to industry shutdowns with obvious negative economic implications. Adding bioenergy to the basket of forest products helps increase silviculture and forest management profitability and improves the economic resilience of other industries.

Australia has one of the highest GHG emissions per capita (UNSTATS, 2014). Coal-fired electricity generation (BREE, 2014) is a significant contributor to these emissions. If Australian government policies shift in favour of substantial reductions in GHG emissions, increased use of woody biomass for energy is an obvious candidate technology. Even without active government-level policy with strong GHG reduction targets and well-defined and achievable roadmaps, some companies and organisations may independently choose to reduce their GHG emissions by using woody biomass. The residential and commercial space heating sector in Canada used 1366 PJ of energy in 2012 (NRCAN, 2014) with no energy from wood reported in the commercial sector and only 169 PJ (12%) of energy from wood in the residential sector. Increased use of biomass thermal and CHP will decrease the quantity of GHG produced from fuel oil and diesel combustion. Such a transition may be attractive for individual communities or industries wanting to reduce their environmental footprint.

SUMMARY AND CONCLUSIONS

As new bioenergy projects are established in countries that are in the early stages of bioenergy development, there are significant opportunities to drive efficient market development through strong technology learning approaches, even for smaller niche applications. Richards et al. (2012) concluded that technological and political barriers to deploying renewables are significant due to substantial knowledge gaps and acceptance of the status quo. These barriers will not be removed with increased investment alone; instead, lessons from other jurisdictions, where policies and alternative technologies have already been tested, may assist in the acceptance of renewable energy. In evaluating the technology learning gains for Swedish bioenergy, Junginger et al. (2005) noted that regions with a large potential for wood-based bioenergy could avoid some obstacles and challenges by capitalising on the experience gained in countries with mature bioenergy industries, such as Sweden and Finland. Where key efficiency drivers can be identified in already well-established biomass supply chains, countries such as Australia and Canada could avoid inefficiencies that may arise during initial supply chain development. Rather than going through the gradual process of developing efficient practices through adaptive management, as identified by Björheden (2006), new supply chains could incorporate efficient approaches from the start.

Although significant benefits can be gained by learning from others, every country and region has unique features that can only be addressed by learning-through-doing. In the interests of mitigating the risk of establishing new supply chains, even those countries with the most widely established bioenergy industries started with relatively small projects. Technology learning-through-doing in niche applications, leveraged with knowledge from the best practices and experience of well-established bioenergy supply chains, provides the best opportunity to set strong foundations for bioenergy development in countries with immature bioenergy industries.

REFERENCES

AANDC-NRCAN, 2011. Status of Off-Grid/Remote Communities in Canada: Aboriginal Affairs and Northern Development Canada and Natural Resources Canada, Government of Canada. Available from: http://www.nrcan.gc.ca/sites/www.nrcan.gc.ca/files/canmetenergy/files/pubs/2013-118_en.pdf

ABS, 2014. 6401.0—Consumer Price Index, Australia, 2014. Available from: http://www.abs.gov.au/AUSSTATS/abs@.nsf/DetailsPage/6401.0Dec%202014?OpenDocument

AEMO, 2013. 100 Per Cent Renewables Study—Modelling Outcomes: Australian Energy Market Operator. Available from: https://www.environment.gov.au/system/files/resources/d67797b7-d563-427f-84eb-c3bb69e34073/files/100-percent-renewables-study-modelling-outcomes-report.pdf

Arriaga, M., Cañizares, C.A., Kazerani, M., 2013. Renewable energy alternatives for remote communities in Northern Ontario, Canada. IEEE Trans. Sustain. Energy 4 (3), 661–670.

Badcock, J., Lenzen, M., 2010. Subsidies for electricity-generating technologies: a review. Energy Policy 38 (9), 5038–5047.

BCBE, 2011. British Columbia Bio-Economy. Government of British Columbia, Vancouver, Canada. Available from: http://www.gov.bc.ca/jtst/down/bio_economy_report_final.pdf.

BEL, 2016. Pathways to the bioenergy village (Wege zum Bioenergiedorf.). Federal Ministry of Food and Agriculture, Germany. Available from: http://www.wege-zum-bioenergiedorf.de (In German).

Björheden, R., 2006. Drivers behind the development of forest energy in Sweden. Biomass Bioenergy 30 (4), 289–295.

BMU, 2011. New Thinking—New Energy: Energy Policy Road Map 2020. Federal Ministry for the Environment, Nature Conservation and Nuclear Safety. Public Relations Division, Government of Germany, Berlin, Germany, Available from: http://www.germany.info/contentblob/2293472/Daten/426258/Roadmap_DD.pdf.

BP, 2013. BP statistical review of world energy. Available from: http://www.bp.com/en/global/corporate/energy-economics/statistical-review-of-world-energy.html

BREE, 2012. Australian Energy Technology Assessment. Bureau of Resources and Energy Economics, Government of Australia, Canberra, Available from: http://www.industry.gov.au/Office-of-the-Chief-Economist/Publications/Documents/aeta/australian_energy_technology_assessment.pdf.

BREE, 2014. Australian Energy Statistics. Bureau of Resources and Energy Economics, Government of Australia, Canberra, Available from: http://www.industry.gov.au/Office-of-the-Chief-Economist/Publications/Documents/aes/2014-australian-energy-statistics.pdf.

Boileau, M.C., 2015. A mine heated with biomass (Une mine chauffée à la biomasse.). Le Monde forestier [online]. Available from: http://www.lemondeforestier.ca/une-mine-chauffee-a-la-biomasse-forestiere/ (In French).

Bürer, M.J., Wüstenhagen, R., 2009. Which renewable energy policy is a venture capitalist's best friend? Empirical evidence from a survey of international cleantech investors. Energ. Policy 37 (12), 4997–5006.

CEER, 2015. Status Review of Renewable and Energy Efficiency Support Schemes in Europe in 2012 and 2013. Document number: C14-SDE-44-03. Brussels, Belgium. p. 67

CER, 2015. Clean Energy Regulator Renewable Energy Target. Small Scale Installations by Postcode. Available from: http://ret.cleanenergyregulator.gov.au/REC-Registry/Data-reports-Smallscale-installations-by-installation-year.

CIEEDAC, 2014. District Energy Inventory For Canada, 2013. Canadian Industrial Energy End-use Data and Analysis Centre, Simon Fraser University, Burnaby, BC, Available from: http://cieedac.sfu.ca/media/publications/District_Energy_Inventory_FINAL_REPORT.pdf.

Dessureault, Y, 2015. The forest sector in Côte-Nord (Qc); How can it develop? (In French: État de la filière forestière de la Côte-Nord: Comment peut-elle se développer?), Symposium: La filière forestière à l'heure des choix, Baie-Comeau, Canada, 12 March, 2015.

Eaton, W.M., Gasteyer, S.P., Busch, L., 2014. Bioenergy futures: Framing sociotechnical imaginaries in local places. Rural Sociol. 79 (2), 227–256.

ECOFYS, 2014. Subsidies and Costs of EU Energy. The European Commission, Brussels, Belgium, Available from: https://ec.europa.eu/energy/sites/ener/files/documents/ECOFYS2014SubsidiesandcostsofEUenergy_11_Nov.pdf.

EEA, 2014. Energy Support Measures and Their Impact on Innovation in the Renewable Energy Sector in Europe. European Environment Agency, Copenhagen, Denmark, (Technical report No 21/2014). Available from: http://www.eea.europa.eu/publications/energy-support-measures.

Egger, C., Öhlinger, C., 2009. Electricity Efficiency Policy in Upper Austria. Tenth International Association of Energy Economics European Conference: Energy, Policies and Technologies for Sustainable Economies, Vienna, Austria, 7–10 September, 2009.

Egger, C., Öhlinger, C., Auinger, B., Brandstätter, B., Richler, N., Dell, G., 2010. Biomass Heating in Upper Austria: Green Energy, Green Jobs., Linz, Austria. Available from: https://www.biomassthermal.org/resource/PDFs/Austria_Biomass_heating_2010.pdf.

FAO, 2010. Global Forest Resources Assessment 2010 Main Report FAO Forestry Paper 163. Food and Agriculture Organization of the United Nations, Rome, Italy, Available from: http://www.fao.org/docrep/013/i1757e/i1757e.pdf.

Government of Australia, 2000. Renewable Energy (Electricity) Act 2000. Government of Australia, Canberra.

Government of Australia, 2001. Renewable Energy (Electricity) Regulations 2001. Government of Australia, Canberra.

Government of Australia, 2011. Renewable Energy (Electricity) Regulations 2001. Regulations as Amended, Taking Into Account Amendments up to Renewable Energy (Electricity) Amendment Regulations 2011 (No. 5). Government of Australia, Canberra.

Government of Australia, 2015. Renewable Energy (Electricity) Amendment Bill 2015. Government of Australia, Canberra.

Haas, R., Panzer, C., Resch, G., Ragwitz, M., Reece, G., Held, A., 2011. A historical review of promotion strategies for electricity from renewable energy sources in EU countries. Renew. Sustain. Energy Rev. 15 (2), 1003–1034.

Hillring, B., 2002. Rural development and bioenergy—experiences from 20 years of development in Sweden. Biomass Bioenergy 23 (6), 443–451.

Hirschl, B., Aretz, A., Prahl, A., Böther, T., Heinbach, K., Pick, D., Funcke, S., 2010. Municipal Value Creation From Renewable Energies (In German: Kommunale Wertschöpfung Durch Erneuerbare Energien). Institute for Ecological Economy Research in cooperation with the Centre for Renewable Energies of the Albert-Ludwig University of Freiburg, Freiburg, Germany, Available from: http://www.mittelstandsfreundliche-kommunen.de/infothek/content/ioew(2010)_kommunale_wertschoepfung_durch_EE.pdf.

Hoffmann, D., 2009. Creation of regional added value by regional bioenergy resources. Renew Sustain. Energy Rev. 13 (9), 2419–2429.

IEA, 2014a. Renewables Information 2014. International Energy Agency, Paris, France.

IEA, 2014b. Electricity Information 2014. International Energy Agency, Paris, France.

IEA, 2014c. Energy Balances of OECD Countries. 2014 Edition. International Energy Agency, Paris, France.

IEA, 2014d. Energy Prices and Taxes. Quarterly Statistics. Fourth Quarter 2014. International Energy Agency, Paris, France.

IEA, 2014e. CO2 Emissions From Fuel Combustion 2014. International Energy Agency, Paris, France, Available from: https://www.iea.org/publications/freepublications/publication/CO-2EmissionsFromFuelCombustionHighlights2014.pdf.

IRENA, 2016. Global Atlas for Renewable Energy: International Renewable Energy Agency. Available from: http://globalatlas.irena.org/default.aspx.

Junginger, M., Faaij, A., Björheden, R., Turkenburg, W., 2005. Technological learning and cost reductions in wood fuel supply chains in Sweden. Biomass Bioenergy 29 (6), 399–418.

Kedron, P., 2015. Environmental governance and shifts in Canadian biofuel production and innovation. Prof. Geogr. 67 (3), 385–395.

Krupa, J., 2012. Identifying barriers to aboriginal renewable energy deployment in Canada. Energy Policy 42, 710–714.

Le Coq, C., Paltseva, E., 2009. Measuring the security of external energy supply in the European Union. Energy Policy 37 (11), 4474–4481.

Moshkelani, M., Marinova, M., Perrier, M., Paris, J., 2013. The forest biorefinery and its implementation in the pulp and paper industry: energy overview. Appl. Therm. Eng. 50 (2), 1427–1436.

NEB, 2013. Canada's Energy Future 2013. Energy Supply and Demand Projections to 2035. National Energy Board, Government of Canada, Ottawa, Canada. Available from: https://www.neb-one.gc.ca/nrg/ntgrtd/ftr/2013/2013nrgftr-eng.pdf.

NRCAN, 2011. The state of Canada's forests—Annual Report 2011. Canadian Forest Service, Ottawa, Canada, Available from: http://cfs.nrcan.gc.ca/publications?id=32683.

NRCAN, 2014. National Energy Use Database. Natural Resources Canada, Ottawa, Canada, Available from: http://oee.nrcan.gc.ca/corporate/statistics/neud/dpa/data_e/databases.cfm.

NSWGOV, 2013. Protection of the Environment Operations (General) Amendment (Native Forest Bio-Material) Regulation. Government of New South Wales, Sydney, NSW.

Obernberger, T., Thek, G., 2008. Cost assessment of selected decentralised CHP application based on biomass combustion and biomass gasification. Sixteenth European Biomass Conference and Exhibition, ETA-Renewable Energies, Valencia.

OECD, 2010. Uranium 2009 resources, production and demand. Joint Report by the OECD Nuclear Energy Agency and the International Atomic Energy Agency, Paris, France. NEA No. 6891. Available from: https://www.oecd-nea.org/ndd/pubs/2010/6891-uranium-2009.pdf.

Painuly, J.P., 2001. Barriers to renewable energy penetration; a framework for analysis. Renew. Energy 24 (1), 73–89.

Paleologou, M., Radiotis, T., Kouisni, L., Jemaa, N., Mahmood, T., Browne, T., Singbeil, D., 2011. New and emerging biorefinery technologies and products for the Canadian forest industry. J. Sci. Technol. Forest Prod. Proc. 1 (3), 6.

Richards, G., Noble, B., Belcher, K., 2012. Barriers to renewable energy development: a case study of large-scale wind energy in Saskatchewan, Canada. Energy Policy 42, 691–698.

Stokes, L.C., 2013. The politics of renewable energy policies: the case of feed-in tariffs in Ontario, Canada. Energy Policy 56, 490–500.

The European Commission, 2014. Communication From the Commission to the European Parliament and the Council. European Energy Security Strategy. COM 2014 330 Final. The European Commission, Brussels, Belgium.

Thrän, D., Fritsche, U., Hennig, C., Rensberg, N., Krautz, A., 2012. Country Report: Germany 2011. IEA Bioenergy Task 40, Leipzig/Darmstadt, Germany, Available from: http://www.bio-energytrade.org/downloads/iea-task-40-country-report-2011-germany.pdf.

Thrän, D., Hennig, C., Thiffault, E., Heinimö, J., Andrade, O., 2014. Development of bioenergy trade in four different settings–the role of potential and policies. Int. Bioenergy Trade, 65–101.

UNFCCC, 2014. Time Series—Annex 1. GHG total excluding LULUCF'. Available from: http://unfccc.int/ghg_data/ghg_data_unfccc/time_series_annex_i/items/3814.php.

UNSTATS, 2014. Environmental Indicators: GHGs. Available from: http://unstats.un.org/unsd/ENVIRONMENT/air_greenhouse_emissions.htm.,

VICGOV, 2013. Energy Legislation Amendment (Feed-in Tariffs and Other Matters) Act 2013. Government of Victoria, Australia, Melbourne, VIC.

Chapter 10

Challenges and Opportunities for the Mobilisation of Forest Bioenergy in the Boreal and Temperate Biomes

Evelyne Thiffault*, Göran Berndes†, Patrick Lamers**

*Department of wood and forest sciences and Research Centre on Renewable Materials, Laval University, Quebec City, Canada; †Physical Resource Theory, Chalmers University of Technology, Gothenburg, Sweden; **Idaho National Laboratory, Idaho Falls, ID, United States of America

Highlights

- Biomass from boreal and temperate forests plays a critical role in meeting global deployment levels for bioenergy, which in turn are necessary to stabilise global GHG emissions and atmospheric carbon levels.
- Compared to other sources of renewable energy (and fossil fuels), bioenergy occupies a unique place due to its multisectoral nature and potential influence on environmental, social and economic conditions at various scales.
- Energy and forest systems can be integrated across landscapes and organisations to establish attractive solutions guided by good governance of forest management and biomass use.
- Substantial gains in forest bioenergy mobilisation can be achieved with an increase in forest management intensity; for some countries, such intensification would likely require a fundamental shift in the forest systems and considerable societal change.
- A multifaceted policy effort to support a broad array of technological, structural, and behavioural changes is needed to enable energy and industrial transitions, as well as the development of supply chains that deliver feedstock to a range of conversion and utilisation routes.

INTRODUCTION

Biomass production in agriculture and forestry will have to increase drastically to support global bioenergy deployment at levels proposed in IEA (2010) and IRENA (2014), for example. As an illustration of possible longer-term biomass demand for energy, the Intergovernmental Panel on Climate Change (Edenhofer et al., 2011) reviewed 164 long-term global energy scenarios and found bio-energy deployment levels in the year 2050 ranging from 80 to 150 EJ/year for 440–600 ppm CO_2^{eq} concentration targets and from 118 to 190 EJ/year for less than 440 ppm CO_2^{eq} concentration targets. To indicate magnitudes, 100 EJ roughly corresponds to 13.7×10^9 m^3 of wood (at 7.3 GJ/m^3 of wood). In com-parison, the total global annual quantity of wood removals only increased from 2.75×10^9 m^3 to around 3×10^9 m^3, from 1990 to 2011 (Köhl et al., 2015). To reach the suggested deployment levels by 2050, significant and harmonised efforts are required across countries. Forest biomass can make an important contribution to the future energy supply, especially in countries with large for-est resources.

The aim of this book was to identify opportunities and challenges for the mobilisation of sustainable forest bioenergy supply chains in the boreal and temperate biomes, in order to support global deployment levels. It focused spe-cifically on countries in North America, the European Union and the temperate forest biome in Oceania. While slower growing than in tropical zones, forests in the studied areas represent an important share of global net primary production (see the chapter Comparison of Forest Biomass Supply Chains from the Boreal and Temperate Biomes and DeLucia et al. (2014)). Also, forest industries utilis-ing the resource (see the chapter Comparison of Forest Biomass Supply Chains from the Boreal and Temperate Biomes) are well-positioned to increase the for-est biomass use for energy.

As seen in the previous chapters, forest biomass supply chains for bioenergy production can take on many forms, with variations in biomass source and man-agement, end-products, conversion technology, logistics, environmental impacts, and markets. Under the appropriate conditions, increasing bioenergy's share in the energy mix can contribute to important global targets, such as reducing GHG emissions, enhancing energy security and promoting sustainable development. However, bioenergy will bring value to society only if the benefits it provides exceed related externalities, as well as the opportunity costs of its development. Compared to other sources of renewable energy (and fossil fuels), bioenergy oc-cupies a unique place due to its multisectoral nature and potential influence on environmental, social and economic conditions at various scales (Kautto, 2010).

Lauri et al. (2014) is an example of studies that derive global supply-cost curves for woody biomass and illustrate the possible magnitude of future sup-plies of different types of forest biomass (Fig. 10.1). As already experienced in several countries, readily accessible forest industry by-products such as bark, black liquor, sawdust and shavings are initially used as bioenergy feedstocks and (as described in the chapter Environmental Sustainability Aspects of Forest Biomass Mobilisation) there are few environmental concerns associated with this biomass use for energy. Logging residues, non-commercial roundwood, and plantations represent complementary resources that can support ramping up to significantly larger scales, if wood energy prices stimulate mobilisation.

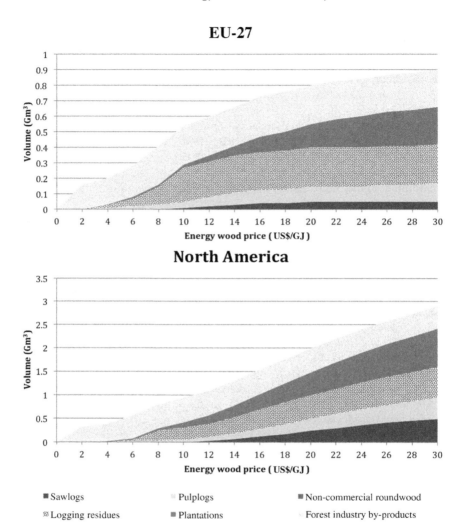

FIGURE 10.1   **Regional wood fibre uses in 2050 at different hypothetical energy wood prices.** From Lauri et al. (2014). Note: Pulplogs include pulplogs and other industrial roundwood. Forest industry by-products include bark, black liquor, sawdust, sawchips and recycled wood. Logging residues consist of harvest losses, branches and stumps. EU-27 includes the European Union and North America includes Canada and the United States.

The impacts of utilising these resources are location specific and their availability across forest landscapes depends on both logistic factors and management/harvest guidelines safeguarding soils, water quality, biodiversity and other values. Other biomass resources such as pulpwood-quality logs may also become used for energy, depending on the competitiveness of bioenergy, compared to other feedstock uses such as sawnwood and pulp and paper production.

Adjustments in forest management and harvesting regimes, due to bioenergy demand, will reflect a range of environmental, social and economic factors, including forest type, climate, legislation, other forest product markets, forest ownership and the character and product portfolio of the associated forest industry. For example, some studies (Buongiorno et al., 2011; Moiseyev et al., 2011; Raunikar et al., 2010; Schwarzbauer and Stern, 2010; Susaeta et al., 2009) report that rising demand for bioenergy causes competition for low-quality sawtimber and pulpwood, while other studies (Cintas et al., 2015; Holmström et al., 2012) find that adaptation of forest management planning to obtain an economically optimal output of forest products, placing equal weight on sawtimber, pulpwood and bioenergy feedstocks, can result in that both pulpwood and forest fuel output increase, due to increased thinning frequency providing both pulpwood and bioenergy.

Thus, bioenergy demand can affect wood use in conventional forest products in antagonistic ways, for example when competition for the same feedstock drives up prices, impairing the competitiveness of conventional forest products. But it can also affect wood use in synergetic ways, where new opportunities and additional incomes from bioenergy use strengthens the forest industry. Besides the stated example where previously unused biomass (such as harvesting or processing residues) is used as bioenergy feedstock, synergies can also be realised through a change in forest management, when considering bioenergy market demand, as described previously. At lower wood energy prices, the benefits from waste and residue use for bioenergy dominate, but feedstock competition naturally becomes stronger as energy wood prices increase (Lauri et al., 2014).

On the other hand, higher prices also provide incentives for expanding forests and improving forest productivity to support increased wood output. For example, Abt et al. (2014) have modelled price-supply interactions in the Southeastern United States, suggesting that the anticipation of emerging bioenergy markets can cause forest expansion (or reduce the rate of forest conversion to other land use, such as urbanisation). As prices increase, forest management also intensifies and the amount of unmanaged forests being put under management rises, as illustrated in Fig. 10.2. The character of the supply side response depends on price levels, but also on the pace of biomass demand growth. Forest planting and silvicultural measures to enhance forest growth represent slower supply side responses than the options to collect felling residues and utilise byflows in the forest industry.

According to data in Raunikar et al. (2010), global industrial roundwood prices have stayed close to or below 100 US\$/m^3 (in real US\$ of 1997) and energy wood prices around 50 US\$/m^3 (6.90 US\$/GJ) for most of the last 50 years. The energy wood price increase of up to 216 US\$/m (30 US\$/GJ) as modelled in Lauri et al. (2014), when driven by an increased demand for energy wood, would imply a significant structural change for the forest sector. However, coal prices as of 2015 are around 3 US\$/GJ and they are expected to stay more or less constant, at least for the near future (EIU-GFS, 2015). It is difficult to

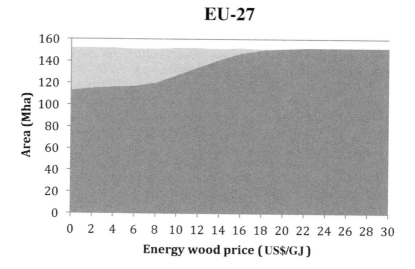

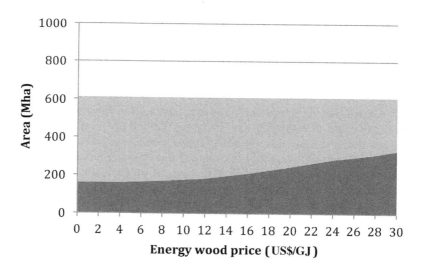

■ Managed forest   ▩ Unmanaged forest

FIGURE 10.2   Regional managed and unmanaged forest areas in 2050, at different energy wood prices. *(From Lauri et al. (2014).)*

imagine that where wood and coal compete for the same market, such as the energy sector (heat and power), the use of woody biomass becomes more attractive – unless the cost of using coal increases drastically. Thus, an increased bioenergy demand would require either subsidies for bioenergy, or policies that reflect the true costs of using fossil fuels (Lauri et al., 2014).

Hence, the most obvious driver to an increased mobilisation and use of forest biomass for energy is an ambitious policy framework. However, as shown in previous chapters and discussed further, complementary and alternative strategies to such policies include identifying barriers, opportunities and solutions to mobilise feedstock more efficiently and promote sustainable supply chains.

CHALLENGES

Variability of Supply Chains

The variation in biomass origins, logistical structures and end-use markets is a barrier in itself, because the very diversity and flexibility associated with bioenergy presents a significant institutional challenge to an emerging industry (Peck et al., 2011):

- While some bioenergy applications are well established (eg, heat and power), others are recently established or not yet commercially available (eg, advanced biofuels);
- There are a wide range of pedo-climatic contexts and scales of production;
- Some logistic systems are 'standard practice' or well developed, while others are emerging from concept stage;
- End-user markets vary from large-scale utilities and international corporations, down to domestic-dwelling scale;
- Feedstock sources can be residue and waste streams, or dedicated crops – or a combination;
- For some chains, bioenergy utilisation constitutes a new form of competition for material or land, or both – for others, markets are new and relatively competition-free;
- Some forest biomass feedstocks are subject to environmental concerns and/or uncertainty concerning benefits (eg, roundwood), while others remain mostly free of controversy (eg, sawmill residues).

Policy

On the policy side, uncertainty still prevails over the future development of forest bioenergy. As shown in the chapters Challenges and Opportunities for the Conversion Technologies Used to Make Forest Bioenergy, Challenges and Opportunities for International Trade in Forest Biomass, as well as Constraints and Success Factors for Woody Biomass Energy Systems in Two Countries with Minimal Bioenergy Sectors, agreements and policies that include instruments such as carbon taxes or renewable energy mandates have had a large effect on the global mobilisation and supply chain development of forest bioenergy (Lamers et al., 2012). Future policies, regulations and other governance will be decisive in creating an attractive investment climate, allowing the energy and other sectors (eg, chemicals) to innovate and invest in the establishment of forest biomass value chains. Increasing globally mobilised forest resources require operational risk reduction, including policy stability, for example adequate and transparent sustainability criteria that can be applied in a predictable way. Clearly, policies will need to adapt over time, in response to technology developments, new knowledge and changing market conditions. But policy changes should neither be abrupt, nor continuously evolving. Without a stable respective policy framework, investments will lack and bioenergy supply (and thus mobilisation) capacities will not be realised.

Logistics

As described in the chapter Challenges and Opportunities of Logistics and Economics of Forest Biomass, a key defining factor for supply chains is economic viability, which in turn is influenced by business risk. Uncertainties and risks on the supply side, with respect to feedstock logistics, are dominated by supply costs and variability in important feedstock properties (eg, moisture content, heating value, bulk density, ash content). Achieving stable operation and high biomass conversion efficiencies requires the management of feedstock variability. Thus, biomass logistics need to address the chemical and physical properties of biomass (Kenney et al., 2013). With forest biomass end-use extending from primarily heat and/or electricity production to a broader range of conversion technologies for energy and other bio-based products, feedstock quality management will need to be adapted to meet requirements of specific conversion processes (INL, 2014). Technology advancement may improve robustness in operations and enable the use of difficult feedstocks, such as coarse organic waste, in some conversion processes. For other conversion processes, stricter requirements for limited biomass variability in quantity, quality and format requires further research and development into effective logistic systems that can meet these requirements.

Also related to logistics, long transport distances and lacking, or poorly organised, markets on the supply and demand sides are commonly perceived bottlenecks to further mobilisation (Athanassiadis et al., 2014). Wide fluctuations in fossil fuel and other competitive products will also be a key factor in economic viability.

Trade

Trade of forest biomass is hampered by regional and national differences in forest production, management, definitions (eg, harvest fractions), respective legislation and enforcement (see the chapter Challenges and Opportunities for International Trade in Forest Biomass). At this point, forest biomass for energy is not subject to trade duties or taxes (unlike liquid biofuels). However, emerging sustainability legislation, including criteria and requirements for supply chain custody – particularly in importing regions, may influence future trade streams. Whether sustainability criteria, such as those outlined in the UK's Timber Procurement Policy, infringe on international agreements under the World Trade Organization (WTO), such as the Technical Barriers to Trade Agreement, is subject to debate (Burrell et al., 2012, Mitchell and Tran, 2009). For the WTO, the concept of sustainability hinges on whether traded goods can be distinguished by how they are made (Process and Production Methods – PPM) or merely by their physical attributes (Goetzl, 2015). The applicability of WTO rules to products differentiated by PPM remains largely unresolved (Goetzl, 2015; Mitchell and Tran, 2009).

Additional trade barriers may lie in a lack of supply chain cooperation/coordination. As suggested by Peck et al. (2011), supply chains of international biomass markets have not yet developed sufficiently so as to function together:

- Emerging chains that do not yet function internally (eg, technology, finance, socio-economic status and performance);
- Chains that do not yet function externally (eg, information and resource exchange with stakeholders/society; collective action/industry organisations, outreach; legitimacy outreach to garner support, acceptance, provide good citizen role; mature interaction with policy-sphere).

Environmental and Social Sustainability

As shown in the chapter Environmental Sustainability Aspects of Forest Biomass Mobilisation, safeguarding environmental sustainability is an important challenge across forest biomass supply chains. Forest biomass procurement may conflict with biodiversity conservation and nature protection goals, as intensified forest management to meet increasing demand bioenergy and other bio-based products may affect ecosystem services negatively (eg, dead wood availability, soil fertility, or forest composition) (Ferranti, 2014). Concerning the contribution of forest bioenergy supply chains to climate change mitigation, the timelag between atmospheric carbon absorption during forest growth and carbon release during biomass combustion leads to GHG mitigation trade-offs between biomass extraction for energy use and the alternative to leave the biomass in the forest – either as forest residues decaying over time, or as living trees that could further sequester carbon (Berndes et al., 2013). Depending on perspective taken – such as spatial (stand versus landscape level) and time scale (short- versus long-term) – different answers are obtained regarding the contribution of forest bioenergy to climate change mitigation (Fig. 10.3).

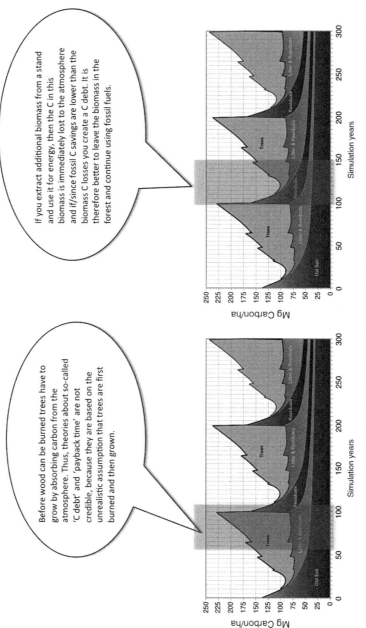

FIGURE 10.3 Diagrams illustrating how stand-level C accounting, with different placement of the accounting window, can lead to contrasting conclusions concerning the GHG mitigation benefit of a forest bioenergy project.

While few would object to the notion that the carbon emitted when forest biomass is used for energy was earlier sequestered from the atmosphere and will be sequestered again if the harvested forest regrows, many object to the previously common approach to disregard the biospheric carbon flows associated with bioenergy systems (carbon neutrality assumption) when quantifying the GHG balance. This is especially so when bioenergy systems are evaluated for their contribution to near-term GHG targets (Ferranti, 2014).

Another aspect of sustainability relates to social factors. As shown in the chapter Economic and Social Barriers Affecting Forest Bioenergy Mobilisation: a Review of the Literature, social acceptability of forest biomass mobilisation should not be taken for granted, since social costs and benefits can be significant. According to Eaton et al. (2014), there is a crucial distinction between mobilising to solve environmental problems and implementing technological development. Even those committed to solving the *environmental problem* of moving beyond fossil fuel-based energy may oppose the *technology* of bioenergy, because of different interpretations and perceptions of costs and benefits: people voice support for transitioning away from conventional and towards renewable sources of energy, but at the same time, attempts to achieve specific technological projects become causes of contest and conflict.

OPPORTUNITIES AND SOLUTIONS

Despite the previous barriers and challenges, the analyses in the previous chapters identified a range of opportunities and solutions promoting mobilisation of forest biomass supply chains in the temperate and boreal biomes.

Technological and Institutional Learning

As mentioned by Jessup and Walkiewicz (2013), the variability of forest biomass supply chains can be seen as a challenge. However, it also offers opportunities for further mobilisation by multiplying the occasions of learning. Junginger et al. (2006) have identified the following mechanisms of learning, which may all play a role in the various contexts of forest biomass mobilisation:

- Learning-by-searching, that is improvements due to Research, Development and Demonstration (RD&D), is the most dominant mechanism in the early stages and, to some extent, also during niche market deployment. Often, also during the stages of pervasive diffusion and saturation, RD&D may contribute to technology enhancements.
- Learning-by-doing comes from the repetition of production process and leads to improvements such as increased labour efficiency, work specialisation and production method improvements.
- Learning-by-using comes from feedback from user experiences and can occur as soon as a technology is being used.

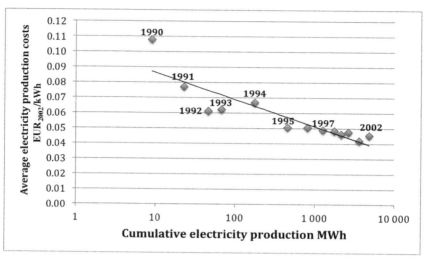

FIGURE 10.4 Experience curve for the average and marginal production cost of electricity from Swedish biofuelled CHP plants from 1990–2002. *(From Junginger et al. (2006).)*

- Learning-by-interacting is related to the increasing diffusion of the technology. During this process, the network interactions between actors, such as research institutes, industry, end-users and policy makers, generally improve and the previously mentioned mechanisms are reinforced.
- Upsizing (or downsizing) a technology may lead to lower specific unit costs (eg, the costs per unit of capacity).

The example from Sweden shows how technological learning has contributed to reduce the costs of electricity production of CHP plants (Fig. 10.4) (Junginger et al., 2006).

An example of the role of technological and institutional learning – and particularly upscaling – is the deployment of the cellulosic biofuel industry in the United States. The current, first-of-a-kind commercial plants rely on a conventional supply chain where corn stover is baled, stored at field-side and delivered in low-density format to the biorefinery. To achieve cost-competitive unit output costs, larger facilities and economies of scale may be required. Large-scale biorefineries (above 5000 tonnes of daily feedstock demand), however, will require a scale-up and transition of the underlying logistics system that includes pre-processing and enables a reduction in feedstock supply risk and variability (Argo et al., 2013; Hess et al., 2009a, 2009b; Muth et al., 2014).

The role of technological and institutional learning for increasing mobilisation was also discussed in the chapter Constraints and Success Factors for Woody Biomass Energy Systems in Two Countries with Minimal Bioenergy Sectors, where it was argued that the ramping up of internal bioenergy demand in Canada could contribute to diversifying biomass end-use markets, while reducing business and investment risk. These could strengthen, stabilise or even expand Canada's international trade with biomass by making supply chains more efficient and cost-competitive.

Work by Röser (2012) has shown that simple technology transfers from one supply chain to the other are not enough to create successful business cases. Technology and know-how need to be combined with existing expertise. Leskinen et al. (2009) described processes of participatory knowledge production in forest bioenergy systems in Finland, in which both expert and local knowledge (ie, tacit knowledge attained by local practitioners through learning-by-doing) were combined, producing mutual learning. An example of this would be economically successful business models that combine calculations of economically and technologically available resources made by experts, local knowledge of practitioners (eg, effective work methods) and social innovations made by local entrepreneurs (eg, forest bioenergy cooperatives) (Leskinen et al., 2009).

Trade

Trade between countries within regions (eg, the EU and North America) and between continents (eg, Canada and the United Kingdom) offers opportunities and incentives for biomass mobilisation. So far, fuelwood, charcoal, wood chips and wood waste have almost exclusively been traded within regions; due to limited homogeneity and bulk density (eg, fuelwood), high moisture content (eg, wood chips), as well as a lack of handling equipment (eg, in transloading stations). Forest biomasss trade for energy between continents has so far only been economically viable for wood pellets, but other alternatives, such as torrefied wood and pyrolysis oil, may become increasingly traded, if commercial production expands. If liquid biofuel production from wood becomes commercially available, it can be expected that these biofuels are traded between continents in the same way as ethanol and biodiesel are traded today. Trade can enable the establishment of logistic systems required for mobilisation of bioenergy supply chains in countries that currently do not have a large domestic market. For example, the current expansion of the US wood pellet production capacity, destined for export to the European Union, could provide a market and logistical 'stepping-stone' to the transition of the US feedstock supply system necessary to support a billion ton national bioeconomy.

Organisation Structures

Another opportunity to supply chain mobilisation that could rise from increased collaboration among stakeholders is the expansion of markets throughout organisation structures working with forest biomass (Athanassiadis et al., 2014; Peck et al., 2011), such as cooperatives, energy firms and trade centres.

The open exchange of information, best practices and market instruments like long-term contracts could be used to improve cooperation between individual forest owners, forest owner organisations, entrepreneurs and forest industry, in order to secure supply and demand. Well-functioning forest owner associations have proved their capability to increase wood supply from small scale private properties. Rural development policies, therefore, should continue to support capacity building of forest owner associations so as to encourage further mobilisation. Furthermore, support for organisation structures such as cooperatives (including items such as the development of professional corporations, associations and formal educational programmes) can also be a way to increase the professionalisation of the workforce in forest biomass supply chains, which has been identified as an important driver for increased biomass mobilisation by Peck et al. (2011). The establishment of a price index for bioenergy products, such as wood pellets, could also contribute to reducing short-term contracts, variable pricing and market instability (EUBIONET2, 2007).

Improvement of Supply Chain Data Reporting

Any strategy for an increased mobilisation of forest biomass requires a relevant assessment of the resource (Kautto, 2010). This is only possible with a sound data base, which is often lacking in the context of predicting forest biomass potential for mobilisation (SFC, 2008). As discussed in the chapter Challenges and Opportunities of Logistics and Economics of Forest Biomass, there is a crucial need to analyse in greater detail the potential biomass supply, taking into account local conditions such as costs, ownerships patterns, quality requirements, infrastructure and environmental considerations. On the other hand, it has been argued that it is mostly the economic viability and sustainable management of bioenergy supply chains that limits the deployment of biomass as a sustainable energy resource, rather than the available quantities of biomass (Siemons et al., 2004).

Related to data gathering is the need for streamlining datasets on traded biomass volumes, across international institutions. This would both improve reporting of trade streams and provide essential information for scientific research on mapping future trade streams, under different policy and potential trade regime scenarios (Lamers et al., 2012).

Integration of Energy and Forest Systems

Management of Biomass Quality

One important step in forest biomass mobilisation is collaboration among stakeholders along the supply chain (Athanassiadis et al., 2014). This includes interactions to get a better understanding of needs in terms of feedstock and end-product characteristics (see the chapter Challenges and Opportunities of Logistics and Economics of Forest Biomass). Technology developers and providers should ensure that the technologies developed are robust enough to handle the variability of forest biomass resources. On the other hand, biomass suppliers need to ensure strict feedstock quality management. Limited or low quality forest fuel causes risks to the downstream processes and unscheduled stoppages lower the profitability of cost-sensitive processes, such as heat production (Laihanen et al., 2013).

Integrated Planning of Bioenergy and Conventional Wood Products

Adequate characterisation and sorting of wood fibre, as early as possible in the supply chain, can provide strategic information that facilitates economic and environmental management decisions on treatments for individual trees and forest stands, improves thinning and harvesting operations and allocates efficiently timber resources for optimal utilisation. For example, the information could be used to sort and grade trees and logs according to their suitability for different end-uses, such as structural products, advanced composites, pulp and paper and bioenergy (Wang, 2012). This should increase the profitability of the entire forest product value chain (including both conventional products, such as sawnwood and paper, as well as bioenergy) and the mobilisation of biomass, as a result of two processes. First, proper identification, inventory and management of biomass for bioenergy (ie, unutilised or unloved fibre by industries of conventional wood products) should increase the total volume of wood harvested per unit area of land and, thus, decrease overall harvesting costs. Second, the addition of bioenergy to the forest product value chain should improve the wood sorting capacity throughout the chain, thus ensuring that only feedstock of suitable quality is processed into each forest product. Optimal allocation of wood resources is vital if wood and bioenergy buyers are to obtain the material most suited to their needs and suppliers are to obtain the best return for their investment in forest land.

An accompanying conclusion to this is that supply costs can be significantly reduced by integrating supplies of wood for both conventional forest products and bioenergy, as described in the chapter Challenges and Opportunities of Logistics and Economics of Forest Biomass, rather than providing them via separate supply chains (Kong et al., 2012). Fibre terminals where wood can be sorted into multiple feedstock assortments and pre-processed (or blended), based on its characteristics, can play a key role in the provision of such flexibility and links back to the previously mentioned organisation and logistic structures. For biomass producers, terminals could also ensure that forest machinery can be utilised effectively year-round. Since raw forest biomass cannot be transported long distances due to its relatively low value, robust value-upgrading at terminals close to the feedstock sources, before long-distance transportation, could be considered (Laihanen et al., 2013).

The forest and the energy sectors use different types of measurements, which hinder communication and coordination between both sectors and market development. Forest industry and energy producers should work jointly on the interoperability of specifications and measures (volumetric and energetic) for wood fuel and wood fuel products, as well as common terminology and conversion factors related to wood for energy (SFC, 2008).

Conversion Efficiency and Cascading Use

Integration of forest and energy systems can help improve the efficiency in biomass resource use. As highlighted in the chapter Challenges and Opportunities for the Conversion Technologies Used to Make Forest Bioenergy, a key is to strive for high efficiency in converting primary biomass into energy products and further into energy services. Higher value forest biomass resources can be used as feedstock several times through cascading systems (Sikkema et al., 2014) where the biomass is processed into a material product, possibly multiple times (eg, construction wood followed by other material applications such as biochemicals and biomaterials), before it is finally used for energy. Reporting, monitoring and research on conversion efficiency and cascading strategies, as well as associated policy development, are important steps towards increased and improved use of forest biomass resources (Kautto, 2010).

Integrated Forest Land Planning for Energy, Conventional Wood Products and Ecosystem Services

There are many different silvicultural practices that could be modified/enhanced in such a way as to incorporate the future forest biomass market considerations at earlier planning stages (SFC, 2008). Along the same lines, forest management approaches aimed at producing forest bioenergy along with conventional forest products should focus on strengthening existing environmental synergies with other forest functions (Ferranti, 2014), as described in the chapter Environmental Sustainability Aspects of Forest Biomass Mobilisation. These synergies could include, for example, forest fire protection, conservation of a balance in soil nutrients and support of biodiversity and water quality. This can be accomplished by increasing biomass removals in areas at high risk of forest fires, in order to reduce fuel loading; by avoiding biomass procurement in sensitive forest areas characterised by poor soils; by leaving adequate amounts of dead wood and trees with cavities in forests to support biodiversity; by avoiding soil compaction, carrying out energy wood extraction at the same time as other forestry operations (Ferranti, 2014). As highlighted in the chapter Environmental Sustainability Aspects of Forest Biomass Mobilisation, a transition to an energy system devoid of fossil fuels and based on renewable sources (which may require short-term increases in GHG emissions) necessitates the adaptation of the forestry sector for a future situation where it is expected to provide ecosystem services, biomaterials and bioenergy.

Development of a Shared Vision

Recognition of Different Views and Understandings

While specific parts of the forest bioenergy supply chain may have the potential to expand rapidly, constraints related to social acceptability (eg, evidence of activities becoming 'trusted' or 'taken for granted' by stakeholders in the general public and in markets) can slow or prevent their deployment. Such issues need to be recognised and must then be factored into longer term plans for the development of the sector (Peck et al., 2011). An emphasis on bioenergy technologies, rather than on the environmental problems that such technologies are meant to address, may harm broader coalition building by failing to make room for the formation of a collective vision of the future (Eaton et al., 2014).

Söderberg and Eckerberg (2013) have also shown that while there is considerable leverage from environmental arguments in favour of bioenergy policy in general, environmental interest groups may remain sceptical. Different environmental arguments seemingly stand against each other. For example, 'bioenergy for green growth' may seem to conflict with 'bioenergy for climate' as described in the chapter Environmental Sustainability Aspects of Forest Biomass Mobilisation (how green and climate-friendly is bioenergy if biodiversity and forests carbon stocks are threatened from increased biomass use?). The simple detection and formal recognition of different scientific understandings or different views of environment can contribute to produce new knowledge, shed new light on the environmental consequences of bioenergy policy and possibly bring stakeholders closer to a shared vision (Eaton et al., 2014; Söderberg and Eckerberg, 2013).

Development of Common Sustainability Criteria

Whether bioenergy development will be beneficial or detrimental for forests – and for people depending on forests for their livelihoods – will be determined by many factors, including legislation, regulations, standards and incentives for biomass production. Development of sustainability criteria for bioenergy is part of the shared vision (Söderberg and Eckerberg, 2013). Junginger et al. (2011) showed that stakeholders along forest biomass supply chains clearly recognise that there is a need to substantiate the sustainable production of biomass. On the other hand, consensus on what should be considered sustainable production and how this should be implemented, certified and monitored is still elusive (Thiffault et al., 2015). In the end, it will probably depend on whether one (or a few) generally accepted systems will become commonplace and whether these systems are workable and affordable. A dialogue aiming to establish internationally accepted sustainability requirements for bioenergy commodities will create new opportunities for sustainable bioenergy trade (Junginger et al., 2011).

Development of Common Technical Standards

Also part of the need for a shared vision is the drive towards technical standardisation of bioenergy products (Junginger et al., 2011). The development of standards (such as the mandates given by the European Commission to the European Committee for Standardization – CEN) can help remove trade barriers, increase market transparency and contribute to public acceptance (Scarlat et al., 2015). The fact that the major producing regions have already started to compare (and possibly align) their technical standards is a sign that international cooperation may lead to new opportunities for international bioenergy trade.

MOBILISATION OF FOREST BIOENERGY POTENTIAL

The chapter Quantifying Forest Biomass Mobilisation Potential in the Boreal and Temperate Biomes presented calculations on the biomass mobilisation potential for a suite of countries analysed in this book. The increased mobilisation of forest biomass was defined as the result of two processes:

- The intensification of forest management activities, in which forestry would appropriate a larger share of forest ecosystem net primary production (NPP), which was reflected in the NPP-to-Roundwood ratio; and
- The intensification of biomass recovery from silvicultural, harvesting and wood processing operations, in which bioenergy would appropriate a larger share of forestry by-products/residues, which was reflected in the Bioenergy-to-Roundwood ratio.

Table 10.1 summarises projections of woody biomass production for various scenarios of increasing mobilisation, namely:

- Scenario 1: an increase of the Bioenergy-to-Roundwood ratio to a minimum of 50% for all studied countries, a ratio that is surpassed by most European countries but well above the ratio observed in important forest countries such as Canada and Russia;
- Scenario 2: an increase of the Bioenergy-to-Roundwood ratio to 83% for all studied countries (the highest ratio observed among the studied countries);
- Scenario 3: an increase of the Roundwood-to-NPP ratio to a minimum of 10% (equivalent to the current average ratio among the studied countries);
- Scenario 4: an increase of the Roundwood-to-NPP ratio to a minimum of 10% and the Bioenergy-to-Roundwood ratio to a minimum of 50%; and
- Scenario 5: an increase of the Roundwood-to-NPP ratio to a minimum of 10% and the Bioenergy-to-Roundwood ratio to 83%.

A comparison of our calculated potentials with targets for bioenergy mobilisation set by international agencies gives insights into the importance of forest biomass supply chains in boreal and temperate biomes. The studied countries could help mobilise up to 19% of the target set for biomass and waste by the IEA in its Blue Map 2050 (146 EJ), from its current level of 3%. The IPCC reported a maximum forest biomass potential of 110 EJ in 2050; current forest biomass production in the studied countries corresponds to 4% of this target and increased mobilisation according to Scenario 5 would correspond to 25%. The Renewable Energy Roadmap (Remap) 2030 of IRENA is based on projections from Smeets et al. (2007) (Table 10.2). Projected values for IRENA's Low scenario in which forest harvesting operations are limited to forest areas already under management are close to our Scenarios 2 and 3, which either involve a larger appropriation of forestry by-products/residues for bioenergy production,

TABLE 10.1 Projections of Forest Biomass Production for Studied Countries (data is in EJ/year)

Country	Production of woody biomass (average 2002–2013)	Scenario 1 Bioenergy: roundwood = 50%	Scenario 2 Bioenergy: roundwood = 83%	Scenario 3 Roundwood: NPP = 10%	Scenario 4 Roundwood: NPP = 10% and bioenergy: roundwood = 50%	Scenario 5 Roundwood: NPP = 10% and bioenergy: roundwood = 83%
Australia	0.19	0.19	0.21	5.63	5.63	6.37
Belgium	0.03	0.03	0.03	0.03	0.03	0.03
Canada	0.48	0.68	1.12	1.91	2.66	4.43
Croatia	0.02	0.02	0.03	0.02	0.02	0.03
Denmark	0.04	0.04	0.04	0.04	0.04	0.04
Finland	0.31	0.31	0.35	0.31	0.31	0.35
Germany	0.38	0.38	0.38	0.38	0.38	0.38
Ireland	0.01	0.01	0.02	0.01	0.01	0.02
New Zealand	0.05	0.09	0.15	0.11	0.22	0.36
Norway	0.05	0.05	0.07	0.10	0.10	0.14
Sweden	0.35	0.35	0.49	0.35	0.35	0.49
United States	2.06	2.06	2.69	3.74	3.74	4.88
Russia	0.14	0.74	1.23	1.19	6.31	10.48
Total all countries	4.09	4.94	6.82	13.80	19.79	28.01
North American countries	2.54	2.74	3.81	5.64	6.40	9.31
European countries	1.31	1.92	2.64	2.41	7.54	11.97
Oceanian countries	0.23	0.28	0.36	5.75	5.85	6.74

Note: North American countries: Canada and the United States. European countries: Belgium, Croatia, Denmark, Finland, Germany, Ireland, Norway, Sweden and Russia. Oceanian countries: Australia and New Zealand.

TABLE 10.2 Projections from Smeets et al. (2007) Used as Basis for IRENA Remap (data is in EJ/year)

	Surplus forest growth	Logging and processing residues	Total
Low scenario (Ecological-economic potential)			
Europe	0	3.60	3.60
North America	0	6.20	6.20
Oceania	0	0.60	0.60
Global	0	14.50	14.50
High scenario (Economic potential)			
Europe	14.00	4.70	18.70
North America	0.20	6.20	6.40
Oceania	0	0.60	0.60
Global	14.60	17.10	31.70

Note: Europe includes East and West Europe, the Baltic States, and the Commonwealth of Independent States (including Russia). The High scenario corresponds to the economic potential, that is the total potential that can be produced at economically profitable levels in the areas of available supply. The Low scenario corresponds to the ecological-economic potential, that is the total biomass based on wood production and utilisation that are limited to forests currently under commercial operation.

or a larger appropriation of forest ecosystem production by the forest sector (both for conventional and bioenergy products). On the other hand, our projections for Scenarios 4 and 5 mobilise 19 and 28 EJ/year, respectively, which is fairly close to the level of IRENA's High scenario; this suggests that achieving such a level of mobilisation would require significant intensification of forest management activities and of biomass recovery from silvicultural, harvesting and wood processing operations.

The analysis also reveals the crucial role played by Russia in global mobilisation of forest bioenergy: to fulfil the country's expected contribution to global targets (ie, up to 10 EJ/year in our scenario 5, from an average of 0.14 EJ/year in the 2002–2013 decade), considerable efforts will be required in institutional learning and capacity-building, over a relatively short timeframe. On the other hand, our projections seem to overestimate grossly the level of forest biomass mobilisation that can occur in New Zealand and Australia, a comparison with IRENA's estimates showing differences of orders of magnitude (Table 10.2). This is likely due to our approach that fails to properly capture the correct NPP of those countries, as explained in the chapter Quantifying Forest Biomass Mobilisation Potential in the Boreal and Temperate Biomes.

Some of the solutions for mobilisation highlighted in this chapter largely address the first of the two suggested processes for increasing mobilisation – that is intensifying biomass recovery from forestry activities. Improving logistics and conversion technologies, increasing quality management of biomass and developing organisation structures supporting stakeholders along the supply chain will make it possible to increase efficiency of practices and extract more energy (and value) from harvested wood, within modernised versions of current forest sector infrastructures and frameworks. Technological learning would play an important role in this modernisation. Such improvements over the next decades would likely allow to reach projections similar to those of Scenario 1 (medium modernisation), or Scenario 2 (important modernisation), with mobilisation of forest biomass from boreal and temperate biomes providing 5–7 EJ/year (Table 10.1).

This is, however, still minuscule compared to the targets set by other agencies. According to our calculations, more substantial gains in mobilisation can only be achieved with an increase in forest management intensity causing a larger appropriation of forest NPP, such as in Scenarios 3 to 5 (14–28 EJ/year). Since sustainability challenges are more likely to arise with intensification of forest management, strong governance schemes and globally accepted sustainability criteria would be important. Moreover, such an intensification would likely require a fundamental shift in the forest systems of many countries. For example, for Canada, reaching a Roundwood-to-NPP ratio of 10% would entail a tripling of the current annual allowable cut (AAC). This would require a drastic change in silviculture practices – currently largely based on extensive forestry – and an opening of yet unmanaged forest areas. The first step in such a momentous shift towards more intensive forestry would undeniably be the development of a shared vision, in which actors would agree on a collective vision about the future national and global forestry and energy systems where bioenergy would occupy a significant place; indeed, a considerable societal change for Canada (see the chapter Constraints and Success Factors for Woody Biomass Energy Systems in Two Countries with Minimal Bioenergy Sectors).

SUMMARY AND CONCLUSIONS

Forest biomass supply chains cover a wide range of biomass sources, logistic systems, conversion technologies, end-products and stakeholders. Several supply chains have proven economically viable and are considered sustainable; others not. We suggest that bioenergy-related policies should be designed in a way that they enhance technological and economic efficiency and environmental sustainability. Pelkonen et al. (2014) suggest the following 'stress-test' for assessing sustainable forest biomass supply chains:

- What is the carbon balance of the processes?
- What are the biodiversity impacts of the processes?
- What are the potential trade-offs (opportunity costs) in terms of forgone alternative forest uses?

- What is the energy efficiency of the process?
- What is the socio-economic viability of the process – that is to what extent is policy support needed and for how long?

Objectives and policies for bioenergy and for renewable energy in general are often formulated and agreed upon at higher decision making levels. However, the design of objectives and policies can be better informed by knowledge and experience at the lower levels of decision making, where the implementation takes place, to integrate renewable energy effectively into the existing energy systems. Local strategies can assist in translating national plans to local level action, while allowing for local level prioritisation and ownership. Local planning can facilitate the identification of the most favourable sites and technologies, improve the understanding of the local environment and its actors and facilitate the integration of policies throughout various sectors, in order to cater for regional ecological complexity (Kautto and Peck, 2012).

Finally, as exemplified in this book, national governments around the world (besides those in the boreal and temperate biomes) have different reasons (policy objectives), as well as different available resources (including climate and available land) for increasing the production and use of forest biomass. Thus, an essential first step in designing appropriate bioenergy policies is to distinguish between the needs and resources of individual countries. Industrialised countries and major exporters of bioenergy among the developing countries should encourage the development of bioenergy where it can be demonstrated that doing so will reduce GHG emissions over the whole life cycle. Other developing countries and those with economies in transition should primarily develop bioenergy to benefit local livelihoods through providing heat and electricity, as well as affordable, safe and more efficient fuels – and so support wider sustainable development goals without jeopardising food security. Ultimately, a multifaceted policy effort to support a broad array of technological, structural and behavioural changes is needed, in order to enable energy and industrial transitions and the development of supply chains that deliver feedstock to a range of conversion and utilisation routes.

REFERENCES

Abt, K.L., Abt, R.C., Galik, C.S., Skog, K.E., 2014. Effect of policies on pellet production and forests in the US South: a technical document supporting the Forest Service update of the 2010 RPA Assessment. US Department of Agriculture Forest Service, Southern Research Station, Asheville, NC, Gen. Tech. Rep. SRS-202. Available at: http://www.srs.fs.usda.gov/pubs/gtr/gtr_srs202.pdf.

Argo, A.M., Tan, E.C.D., Inman, D., Langholtz, M.H., Eaton, L.M., Jacobson, J.J., Wright, C.T., Muth, D.J., Wu, M.M., Chiu, Y.-W., Graham, R.L., 2013. Investigation of biochemical biorefinery sizing and environmental sustainability impacts for conventional bale system and advanced uniform biomass logistics designs. Biofuel. Bioprod. Bioref. 7 (3), 282–302.

Athanassiadis, D., Wallsten, J., Spinelli, R., Rodriguez, J., Raitila, J., te Raa, R., Vos, J., Matthias, D., Turkmengil, T., Walkiewicz, J., 2014. Technological and economic barriers to introduce and apply innovations in forest energy sector–D6.1. Innovative and effective technology and logistics for forest residual biomass supply in the EU. INFRES, Joensuu, Finland, Rapport INFRES- D6.1. Available at: http://www.infres.eu/en/results/.

Berndes, G., Ahlgren, S., Börjesson, P., Cowie, A.L., 2013. Bioenergy and land use change—state of the art. Wiley Interdis. Rev. Energy Environ. 2 (3), 282–303.

Buongiorno, J., Raunikar, R., Zhu, S., 2011. Consequences of increasing bioenergy demand on wood and forests: An application of the Global Forest Products Model. J. Forest Econ. 17 (2), 214–229.

Burrell, A., Gay, S.H., Kavallari, A., 2012. The compatibility of EU biofuel policies with global sustainability and the WTO. World Econ. 35 (6), 784–798.

Cintas, O., Berndes, G., Cowie, A.L., Egnell, G., Holmström, H., Ågren, G.I., 2015. The climate effect of increased forest bioenergy use in Sweden: evaluation at different spatial and temporal scales. Wiley Interdisc. Rev. Energ. Environ.

DeLucia, E.H., Gomez-Casanovas, N., Greenberg, J.A., Hudiburg, T.W., Kantola, I.B., Long, S.P., Miller, A.D., Ort, D.R., Parton, W.J., 2014. The theoretical limit to plant productivity. Environ. Sci. Technol. 48 (16), 9471–9477.

Eaton, W.M., Gasteyer, S.P., Busch, L., 2014. Bioenergy futures: framing sociotechnical imaginaries in local places. Rural Sociol. 79 (2), 227–256.

Edenhofer, O., Pichs-Madruga, R., Sokona, Y., Seyboth, K., Matschoss, P., Kadner, S., Zwickel, T., Eickemeier, P., Hansen, G., Schlömer, S., Stechow, C.V., 2011. IPCC special Report on Renewable Energy Sources and Climate Change Mitigation. Cambridge University Press, Cambridge, UKand New York, NY.

EIU-GFS, 2015. EIU Economic and Commodity Forecast, February 2015. Coal. Global Forecasting Service - Economist Intelligence Unit, London, UK, Available at: http://gfs.eiu.com.

EUBIONET2, 2007. Solid biomass mobilisation for the forest-based industries and the bio-energy sector. Proceeding from a seminar during European Paper Week 2007, Brussels, Belgium.

Ferranti, F., 2014. Energy wood: a challenge for European forests Potentials, environmental implications, policy integration and related conflicts. European Forest Institute, Joensuu, Finland, EFI Technical Report 95, 2014.

Goetzl, A., 2015. Development in the global trade of wood pellets. Washington DC, USA: US International Trade Commission. (ID-039). Available at: http://www.usitc.gov/publications/332/wood_pellets_id-039_final.pdf.

Hess, J.R., Wright, C.T., Kenney, K.L., Searcy, E.M., 2009a. Uniform-format solid feedstock supply system: a commodity-scale design to produce an infrastructure-compatible bulk solid from lignocellulosic biomass. Executive summary. Idaho National Laboratory, Idaho Falls, ID, INL/EXT-09-15423. Available at: https://inldigitallibrary.inl.gov/sti/4408280.pdf.

Hess, J.R., Kenney, K., Wright, C., Perlack, R., Turhollow, A., 2009b. Corn stover availability for biomass conversion: situation analysis. Cellulose 16 (4), 599–619.

Holmström, H., Wikström, P., Eriksson, L., 2012. Energy optimized forest management (In Swedish: Energioptimera skogsbruket). Svensk Fjärrvärme AB, Stockholm, Sweden, Fjärrsyn Rapport 2012:13.

IEA, 2010. Energy Technology Perspectives 2010 Scenarios and Strategies to 2050. Organisation for ECD/IEA, Paris, France. Available at: https://www.iea.org/publications/freepublications/publication/etp2010.pdf

INL, 2014. Feedstock Supply System Design and Economics for Conversion of Lignocellulosic Biomass to Hydrocarbon Fuels. Conversion Pathway: Fast Pyrolysis and Hydrotreating Bio-oil Pathway, the 2017 Design CaseIdaho National Laboratory, Idaho Falls, ID, INL/EXT-14-31211. Available at: https://inldigitallibrary.inl.gov/sti/6038147.pdf.

IRENA, 2014. REmap 2030 A Renewable Energy Roadmap. IRENA, Abu Dhabi, Report available at: http://www.irena.org/remap.

Jessup, E.L., Walkiewicz, J., 2013. Adapted forestry practices for improved biomass recovery. INFRES—Innovative and effective technology and logistics for forest residual biomass supply in the EU, Joensuu, Finland, Available at: http://www.infres.eu/en/results/.

Junginger, M., de Visser, E., Hjort-Gregersen, K., Koornneef, J., Raven, R., Faaij, A., Turkenburg, W., 2006. Technological learning in bioenergy systems. Energy Pol. 34 (18), 4024–4041.

Junginger, M., van Dam, J., Zarrilli, S., Ali Mohamed, F., Marchal, D., Faaij, A., 2011. Opportunities and barriers for international bioenergy trade. Energy Pol. 39 (4), 2028–2042.

Kautto, N., 2010. Planning biomass for energy: examining the why, how and what of sound biomass policy. IIIEE, Lund, Sweden.

Kautto, N., Peck, P., 2012. Regional biomass planning – helping to realise national renewable energy goals? Renew. Energy 46 (0), 23–30.

Kenney, K.L., Smith, W.A., Gresham, G.L., Westover, T.L., 2013. Understanding biomass feedstock variability. Biofuels 4 (1), 111–127.

Köhl, M., Lasco, R., Cifuentes, M., Jonsson, Ö., Korhonen, K.T., Mundhenk, P., de Jesus Navar, J., Stinson, G., 2015. Changes in forest production, biomass and carbon: results from the 2015 UN FAO Global Forest Resource Assessment. Forest Ecol. Manag. 352, 21–34.

Kong, J., Rönnqvist, M., Frisk, M., 2012. Modeling an integrated market for sawlogs, pulpwood, and forest bioenergy. Canadian J. Forest Res. 42 (2), 315–332.

Laihanen, M., Karhunen, A., Ranta, T., 2013. Possibilities and challenges in regional forest biomass utilization. J. Renew. Sustain. Energy 5 (3), 033121.

Lamers, P., Junginger, M., Hamelinck, C., Faaij, A., 2012. Developments in international solid biofuel trade—an analysis of volumes, policies, and market factors. Renew. Sustain. Energy Rev. 16 (5), 3176–3199.

Lauri, P., Havlík, P., Kindermann, G., Forsell, N., Böttcher, H., Obersteiner, M., 2014. Woody biomass energy potential in 2050. Energy Pol. 66, 19–31.

Leskinen, L.A., Sikanen, L., Leskinen, P., Röser, D., Viiri, H., Peltola, T., 2009. Challenges of promoting sustainable forest energy technology and know-how. IOP Conference Series: Earth Environ. Sci. 6 (18.).

Mitchell, A.D., Tran, C., 2009. The consistency of the EU renewable energy directive with the WTO agreements. Georgetown Law Faculty Working Papers. Paper 119. 15 p. Available at: http://scholarship.law.georgetown.edu/fwps_papers/119.

Moiseyev, A., Solberg, B., Kallio, A.M.I., Lindner, M., 2011. An economic analysis of the potential contribution of forest biomass to the EU RES target and its implications for the EU forest industries. J. Forest Econ. 17 (2), 197–213.

Muth, D.J., Langholtz, M.H., Tan, E.C., Jacobson, J.J., Schwab, A., Wu, M.M., Argo, A., Brandt, C.C., Cafferty, K.G., Chiu, Y.W., 2014. Investigation of thermochemical biorefinery sizing and environmental sustainability impacts for conventional supply system and distributed preprocessing supply system designs. Biofuel. Bioprod. Bioref. 8 (4), 545–567.

Peck, P., Berndes, G., Hektor, B., 2011. Mobilising global bioenergy supply chains. Keys to unlocking the potential of bioenergy. Swedish Energy Agency and IIIEE, Lund, Sweden, IIIEE Report 2011:02. Available at: http://lup.lub.lu.se/luur/download?func=downloadFile&recordOId=2369139&fileOId=2369140.

Pelkonen, P., Mustonen, M., Asikainen, A., Egnell, G., Kant, P., Leduc, S., Pettenella, D., 2014. Forest Bioenergy for Europe. European Forest Institute, Joensuu, Finland, Part of the EFI's series what science can tell us.

Raunikar, R., Buongiorno, J., Turner, J.A., Zhu, S., 2010. Global outlook for wood and forests with the bioenergy demand implied by scenarios of the intergovernmental panel on climate change. Forest Pol. Econ. 12 (1), 48–56.

Röser, D., 2012.Operational efficiency of forest energy supply chains in different operational environments. Doctorate, University of Eastern Finland. Available at: http://www.metla.fi/dissertationes/df146.pdf

Scarlat, N., Dallemand, J.-F., Monforti-Ferrario, F., Nita, V., 2015. The role of biomass and bioenergy in a future bioeconomy: policies and facts. Environ. Dev. 15, 3–34.

Schwarzbauer, P., Stern, T., 2010. Energy vs material: Economic impacts of a "wood-for-energy scenario" on the forest-based sector in Austria—a simulation approach. Forest Pol. Econ. 12 (1), 31–38.

SFC, 2008. Mobilisation and efficient use of wood and wood residues for energy generation: Standing Forestry Committee ad hoc Working Group II on mobilisation and efficient use of wood and wood residues for energy generation. Available at: http://ec.europa.eu/agriculture/fore/publi/sfc_wgii_final_report_072008_en.pdf

Siemons, R., Vis, M., van den Berg, D., Mc Chesney, I., Whiteley, M., Nikolaou, N., 2004. Bioenergy's role in the EU energy market: a view of developments until 2020. Report to the European CommissionBTG Biomass Technology Group BV, Enschede, the Netherlands, Available at: https://np-net.pbworks.com/f/BTG+%282004%29+Bioenergy%5C's+role+in+the+EU+2000-2020.pdf.

Sikkema, R., Junginger, M., van Dam, J., Stegeman, G., Durrant, D., Faaij, A., 2014. Legal harvesting, sustainable sourcing and cascaded use of wood for bioenergy: their coverage through existing certification frameworks for sustainable forest management. Forests 5 (9), 2163–2211.

Smeets, E.M., Faaij, A.P., Lewandowski, I.M., Turkenburg, W.C., 2007. A bottom-up assessment and review of global bio-energy potentials to 2050. Prog. Energ. Combust. Sci. 33 (1), 56–106.

Söderberg, C., Eckerberg, K., 2013. Rising policy conflicts in Europe over bioenergy and forestry. Forest Pol. Econ. 33 (0), 112–119.

Susaeta, A., Alavalapati, J.R.R., Carter, D.R., 2009. Modeling impacts of bioenergy markets on non-industrial private forest management in the Southeastern United States. Nat. Res. Model. 22 (3), 345–369.

Thiffault, E., Endres, J., McCubbins, J.S., Junginger, M., Lorente, M., Fritsche, U., Iriarte, L., 2015. Sustainability of forest bioenergy feedstock supply chains: local, national and international policy perspectives. Biofuel. Bioprod. Bioref. 9 (3), 283–292.

Wang, X., 2012. Advanced sorting technologies for optimal wood products and woody biomass utilization. Bio-based Material Science and Engineering Conference, Oct. 21–23, Changsha, China, pp. 175–179.

Abbreviations

AAC	Allowable Annual Cut
ATFS	American Tree Farm System
AU$	Australian dollar
BMP	Best Management Practices
CA$	Canadian dollar
CFD	Contracts for Difference (UK)
CHP	Combined heat and power
CO_2	Carbon dioxide
CSA	Canadian Standards Association
CTL	Cut-to-length
DHS	District heating system
EC	European Commission
EJ	Exajoule = 1×10^{18} joules
EU	European Union
€	Euro
FAO	Food and Agriculture Organization of the United Nations
FOB	Free On-Board (price)
FSC	Forest Stewardship Council
GDP	Gross domestic product
GHG	Greenhouse gas
GJ	Gigajoule = 1×10^9 joules
Gm^3	1×10^9 cubic metres
GWh	Gigawatt-Hours = 1×10^9 watt-hours
ha	Hectare
HHV	High heating value
IEA	International Energy Agency
ILUC	Indirect Land-Use Change
IRENA	International Renewable Energy Agency
IWPB	International Wood Pellet Buyer (Initiative)
ktonnes	Thousand metric tonnes
LCOE	Levellised cost of energy
LHV	Low heating value
LNG	Liquefied natural gas
LPG	Liquefied petroleum gas
LUC	Land-use change
LULUCF	Land-use, land-use change and forestry
m^3	cubic metre
Mha	1×10^6 hectares
Mm^3	1×10^6 cubic metres
MS	Member State (EU)
Mtoe	Megatonne oil equivalent = 1×10^6 tonne oil equivalent
Mtonnes	Million metric tonnes
$MW_{el/th}$	Megawatt electric/thermal (capacity)

MWh	**Megawatt-Hours** = 1×10^6 **watt-hours**
NGO	**Non-governmental organisation**
NPP	**Net primary production**
NREAP	**National Renewable Energy Action Plan**
NW	**North-West**
ODMT	**Oven-dry metric tonne**
PEFC	**Programme for the Endorsement of Forest Certification**
PJ	**Petajoule** = 1×10^{15} **joules**
PPM	**Process and Production Methods**
PV	**Photovoltaics**
REC	**Renewable Energy Certificate (US)**
RED	**Renewable Energy Directive (2009/28/EC)**
RHI	**Renewable Heat Incentive (UK)**
RO	**Renewable Obligation (UK)**
RPS	**Renewable Portfolio Standard (US)**
RTFO	**Renewable Transport Fuel Obligation (UK)**
SBP	**Sustainable Biomass Partnership**
SE	**Southeast**
SFI	**Sustainable Forestry Initiative**
SFM	**Sustainable Forest Management**
SME	**Small and medium size enterprise**
tcm	**Trillion cubic metre (oil and gas)** = 1×10^{12} **cubic metres**
TWh	**Terawatt-Hours** = 1×10^{12} **watt-hours**
UK	**United Kingdom**
US	**United States of America**
US$	**American dollar (United States of America)**
WP	**Wood pellets**
WPe	**Wood pellet equivalent**
WTO	**World Trade Organisation**

Unit Conversion

Prefixes of the Metric System

Prefix	Symbol	Numerical	Exponential
exa	E	1,000,000,000,000,000,000	10^{18}
peta	P	1,000,000,000,000,000	10^{15}
tera	T	1,000,000,000,000	10^{12}
giga	G	1,000,000,000	10^{9}
mega	M	1,000,000	10^{6}
kilo	k	1000	10^{3}

1 MWh = 1 Megawatt-hour = 3.6 gigajoules.
1 Mtoe = 1 Megatonne oil equivalent = 0.0418 exajoule.
1 WPe = 1 Wood pellet equivalent = 17.6 gigajoules per metric tonne.
1 cubic metre (m^3) of solid content of wood chips = 7.3 gigajoules, based on Norway spruce, assuming a specific gravity (solid matter content) of 400 kilograms per cubic metre of solid wood and wood chips with a moisture content of approx. 40%, which is equal to the moisture content in storage-dry wood chips.

Mobilisation of Forest Bioenergy in the Boreal and Temperate Biomes

Author Index

Subject Index

Printed in the United States
By Bookmasters